Web YINGYONG KAIFA JICHU
Web 应用开发基础

主编 谢钟扬 危孟君 董海峰
主审 符开耀 王 雷

西北工业大学出版社

图书在版编目（CIP）数据

Web 应用开发基础/谢钟扬,危孟君,董海峰主编.
—西安:西北工业大学出版社,2016.8(2019.7 重印)
ISBN 978-7-5612-5008-2

Ⅰ.①W… Ⅱ.①谢… ②危… ③董… Ⅲ.①网页制作工具—程序设计 Ⅳ.①TP393.092.2

中国版本图书馆 CIP 数据核字(2016)第 196270 号

出版发行：	西北工业大学出版社
通信地址：	西安市友谊西路 127 号　邮编:710072
电　　话：	(029)88493844　88491757
网　　址：	www.nwpup.com
印 刷 者：	兴平市博闻印务有限公司
开　　本：	787 mm×1 092 mm　1/16
印　　张：	22
字　　数：	534 千字
版　　次：	2016 年 8 月第 1 版　2019 年 7 月第 3 次印刷
定　　价：	55.00 元

前 言

　　Web 程序设计是计算机软件开发的重要领域,有多种流行的开发技术应用于 Web 程序设计领域,其中以基于 Java 的 JSP 技术和基于 C#的 net 技术应用最为广泛。本书以建设工程监管信息系统、电子商务购物网站、建设用地审批电子报盘管理系统、网上书店、Blog 系统、物流管理系统、码头中心船货申报系统、易居房产信息网、教研室管理系统、银行信贷管理系统等实际项目为基础,根据学生由浅入深的习得规律,按照功能模块和岗位知识要求形成单个的开发任务。针对每个任务的完成提供任务描述、功能描述、要求、必备知识、解决思路、操作步骤等内容,指导学生从一个个的实际开发任务去认识和学习 Web 程序设计。

　　本书在编写风格上注重知识、技术的实用性,通过案例强化实践技能,语言力求简洁生动、通俗易懂,主要以 Web 程序设计开发人员的岗位培养目标为核心,紧紧围绕岗位对应的职业能力和职业素质需求,选取具有典型性和代表性的项目,并以其为载体整合、序化教学内容,以实际工作任务为脉络展开教学过程,采用"项目导向、任务驱动"的方式设计课程内容的引入、示范、展开、解决、提高、实训等过程,以"教、学、做"一体化的形式带动学生自主学习。

　　本书可作为高等职业院校、高等专科学校的教材,也可作为 Web 程序设计初学者的自学用书,还可作为从事信息系统开发的设计人员的参考用书。

　　在本书的编写过程中,得到了湖南软件职业学院谭长富院长、符开耀副院长、王雷教授等领导和专家的大力支持与热心帮助,在此表示衷心感谢。

　　本书的出版还得到湖南软件职业学院教学质量工程项目(基于学科交叉的《Web 程序设计(JSP)》课程建设与研究)的部分资助;本书的内容参考了国内外有关单位的研究成果,在此一并表示感谢。

　　另外,由于本书的编写目的定位于 Web 程序设计的基础知识与案例分析相结合,试图让读者在深入了解 Web 程序设计的相关概念与关键技术的基础上,能尝试开展 Web 程序设计的一些初步编程工作,本书的内容编写与结构组织具有一定的难度,加之笔者水平有限,虽然几经修改,但书中仍然会难免存在一些疏漏与不足之处,敬请读者、专家以及同行的批评指正,在此先行表示感谢。

编　者

2016 年 6 月

目　　录

任务一：建设工程监管信息系统报建申请模块 ·· 1
　　一、任务描述 ·· 1
　　二、功能描述 ·· 1
　　三、要求 ·· 2
　　四、必备知识 ·· 3
　　五、解题思路 ·· 4
　　六、操作步骤 ·· 4

任务二：建设工程监管信息系统企业信息模块 ··· 19
　　一、任务描述 ··· 19
　　二、功能描述 ··· 19
　　三、要求 ·· 20
　　四、必备知识 ··· 21
　　五、解题思路 ··· 22
　　六、操作步骤 ··· 22

任务三：建设工程监管信息系统信息列表模块 ··· 36
　　一、任务描述 ··· 36
　　二、功能描述 ··· 36
　　三、要求 ·· 36
　　四、必备知识 ··· 39
　　五、解题思路 ··· 39
　　六、操作步骤 ··· 40

任务四：建设工程监管信息系统交易流程模块 ··· 58
　　一、任务描述 ··· 58
　　二、功能描述 ··· 58
　　三、要求 ·· 59
　　四、必备知识 ··· 61

五、解题思路 ·· 61
　　六、操作步骤 ·· 62

任务五:电子商务购物网站产品查询模块 ·· 84
　　一、任务描述 ·· 84
　　二、功能描述 ·· 84
　　三、要求 ·· 85
　　四、必备知识 ·· 86
　　五、解题思路 ·· 86
　　六、操作步骤 ·· 87

任务六:建设用地审批电子报盘管理系统行政区划模块 ·· 94
　　一、任务描述 ·· 94
　　二、功能描述 ·· 94
　　三、要求 ·· 96
　　四、必备知识 ·· 96
　　五、解题思路 ·· 97
　　六、操作步骤 ·· 97

任务七:建设用地审批电子报盘管理系统补偿标准模块 ·· 121
　　一、任务描述 ··· 121
　　二、功能描述 ··· 121
　　三、要求 ·· 122
　　四、必备知识 ··· 123
　　五、解题思路 ··· 124
　　六、操作步骤 ··· 124

任务八:建设用地审批电子报盘管理系统审批模块 ·· 140
　　一、任务描述 ··· 140
　　二、功能描述 ··· 140
　　三、要求 ·· 141
　　四、必备知识 ··· 142
　　五、解题思路 ··· 143
　　六、操作步骤 ··· 143

任务九:建设用地审批电子报盘管理系统供地方案模块 ·· 157
　　一、任务描述 ··· 157
　　二、功能描述 ··· 157
　　三、要求 ·· 158

四、必备知识 159
　　五、解题思路 160
　　六、操作步骤 160

任务十：网上书店图书信息模块 178
　　一、任务描述 178
　　二、功能描述 178
　　三、要求 179
　　四、必备知识 181
　　五、解题思路 181
　　六、操作步骤 182

任务十一：Blog 系统日志信息模块 217
　　一、任务描述 217
　　二、功能描述 217
　　三、要求 218
　　四、必备知识 218
　　五、解题思路 219
　　六、操作步骤 219

任务十二：物流管理系统的公司信息管理 249
　　一、任务描述 249
　　二、功能描述 249
　　三、要求 250
　　四、必备知识 252
　　五、解题思路 256
　　六、操作步骤 257

任务十三：物流管理系统的客户信息管理 270
　　一、任务描述 270
　　二、功能描述 270
　　三、要求 271
　　四、必备知识 273
　　五、解题思路 276
　　六、操作步骤 277

任务十四：物流管理系统的车辆类型管理 279
　　一、任务描述 279
　　二、功能描述 279

三、要求 ··· 280
　　四、必备知识 ·· 282
　　五、解题思路 ·· 282
　　六、操作步骤 ·· 282

任务十五：物流管理系统的车队信息管理 ·· 285
　　一、任务描述 ·· 285
　　二、功能描述 ·· 285
　　三、要求 ··· 286
　　四、必备知识 ·· 287
　　五、解题思路 ·· 287
　　六、操作步骤 ·· 288

任务十六：码头中心船货申报系统的船货作业信息管理 ·························· 290
　　一、任务描述 ·· 290
　　二、功能描述 ·· 290
　　三、要求 ··· 291
　　四、必备知识 ·· 292
　　五、解题思路 ·· 292
　　六、操作步骤 ·· 293

任务十七：码头中心船货申报系统的进出港旅客流量信息管理 ··················· 295
　　一、任务描述 ·· 295
　　二、功能描述 ·· 295
　　三、要求 ··· 296
　　四、必备知识 ·· 298
　　五、解题思路 ·· 298
　　六、操作步骤 ·· 298

任务十八：易居房产信息网的楼盘信息管理系统报建申请模块 ··················· 300
　　一、任务描述 ·· 300
　　二、功能描述 ·· 300
　　三、要求 ··· 301
　　四、必备知识 ·· 303
　　五、解题思路 ·· 303
　　六、操作步骤 ·· 303

任务十九：易居房产信息网的会员管理 ·· 306
　　一、任务描述 ·· 306

二、功能描述 ··· 306
　　三、要求 ··· 307
　　四、必备知识 ··· 309
　　五、解题思路 ··· 310
　　六、操作步骤 ··· 310

任务二十：易居房产信息网的房产出租信息管理 ·· 312
　　一、任务描述 ··· 312
　　二、功能描述 ··· 312
　　三、要求 ··· 313
　　四、必备知识 ··· 315
　　五、解题思路 ··· 316
　　六、操作步骤 ··· 316

任务二十一：教研室管理系统的个人信息管理 ·· 318
　　一、任务描述 ··· 318
　　二、功能描述 ··· 318
　　三、要求 ··· 319
　　四、必备知识 ··· 321
　　五、解题思路 ··· 321
　　六、操作步骤 ··· 322

任务二十二：教研室管理系统的考勤功能 ·· 324
　　一、任务描述 ··· 324
　　二、功能描述 ··· 324
　　三、要求 ··· 325
　　四、必备知识 ··· 327
　　五、解题思路 ··· 327
　　六、操作步骤 ··· 327

任务二十三：银行信贷管理系统的保证金管理 ·· 329
　　一、任务描述 ··· 329
　　二、功能描述 ··· 329
　　三、要求 ··· 330
　　四、必备知识 ··· 332
　　五、解题思路 ··· 332
　　六、操作步骤 ··· 333

任务二十四：银行信贷管理系统的质押信息管理 …………………………………… 335
　一、任务描述 ………………………………………………………………………… 335
　二、功能描述 ………………………………………………………………………… 335
　三、要求 ……………………………………………………………………………… 337
　四、必备知识 ………………………………………………………………………… 338
　五、解题思路 ………………………………………………………………………… 339
　六、操作步骤 ………………………………………………………………………… 339

任务一：建设工程监管信息系统报建申请模块

一、任务描述

你作为《建设工程监管信息系统》项目开发组的程序员，请实现如下功能：
➢ 建设工程项目信息的列表显示；
➢ 建设工程项目信息的添加。

二、功能描述

（1）点击建设工程项目信息列表页面左边导航条中的"建设工程项目施工报建申请"，则在右边的主体部分显示项目信息列表，如图 1.1 所示。

图 1.1　建设工程项目信息列表页面

（2）点击图 1.1 中的"新建工程"按钮，则进入建设工程项目信息录入页面，如图 1.2 所示。

（3）对图 1.2 中打"＊"号的输入部分进行必填并校验。

（4）点击图 1.2 中"确定"按钮，在项目信息表中增加一条项目信息。

（5）项目信息增加成功后，自动定位到建设工程项目信息列表页面，显示更新后的项目信息列表，如图 1.1 所示。

（6）测试程序，通过建设工程项目信息录入页面增加两条以上项目信息。

图 1.2　建设工程项目信息录入页面

三、要求

1. 界面实现

以提供的素材为基础,实现图 1.1、图 1.2 所示页面。

2. 数据库实现

(1)创建数据库 ConstructionDB。

(2)创建项目信息表(T_project),表结构见表 1.1。

表 1.1　项目信息表(T_project)表结构

字段名	字段说明	字段类型	允许为空	备注
Project_id	工程编号	varchar(32)	否	主键
Project_name	工程名称	varchar(64)	否	
Deputy_name	法人代表	varchar(16)	是	
Telephone	电话	varchar(16)	是	
Addr	地址	varchar(64)	是	

(3)在表 T_project 中插入记录,见表 1.2。

表 1.2　项目信息记录

Project_id	Project_name	Deputy_name	Telephone	Addr
2003-01	住宅小区一期工程	张三	2626266	长沙市天心区
2003-02	教学大楼	王平	8374777	长沙市芙蓉区

3. 功能实现

(1)功能需求如图 1.3 所示。

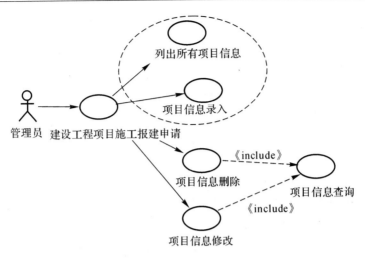

图1.3　建设工程项目施工报建申请模块用例图

(2) 依据项目信息列表活动图完成项目信息列表显示功能,如图1.4所示。

(3) 依据添加项目信息活动图完成添加项目信息功能,如图1.5所示。

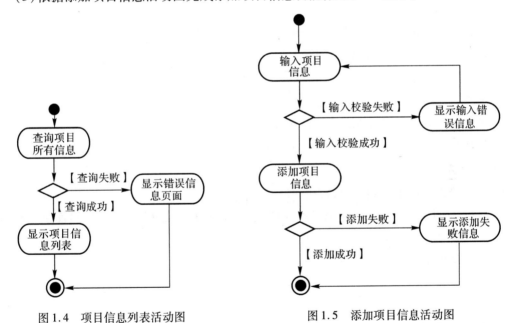

图1.4　项目信息列表活动图　　　　图1.5　添加项目信息活动图

四、必备知识

1. 数据库相关知识

(1) 使用 MS SQL Server 2005/2008 创建数据库,创建数据表,设置表的字段,数据类型,主键,外键,约束。

(2) 向数据表插入、删除、修改、查询数据。

2. 页面相关知识

（1）使用 HTML 制作项目页面。

（2）使用 CSS 控制页面的样式。

（3）使用 JavaScript 对页面必要的内容进行校验。

3. JSP 相关知识

（1）使用 JSTL 标准标签库控制页面显示逻辑。

（2）理解 JSP 的 request,response,session,application 的概念。

（3）使用 EL 表达式在页面显示数据。

（4）合理的使用转发和重定向控制项目的页面跳转。

（5）使用 JDBC 与数据库进行交互。

（6）MVC 模式下的分层架构,控制器,视图,模型的划分和通信。

五、解题思路

1. 数据库思路

（1）根据项目要求创建数据库和数据表,向数据表中插入合适的测试数据。

（2）导入 JDBC 驱动包,编写 JDBC 的连接工具代码。

（3）编写数据操作对象代码,负责进行与数据库的交互操作。

2. 视图层思路

（1）将提供的素材页面改写为 JSP 页面

（2）JSP 使用 JSTL 和 EL 负责控制页面显示的逻辑

3. 控制层思路

（1）使用 Servlet 类控制一次请求响应过程的处理。

（2）由 Servlet 按照顺序进行请求的处理、数据库交互、模型存取和封装、页面跳转逻辑控制等。

4. 模型层思路

使用 JavaBean 作为模型层,封装数据和行为。

六、操作步骤

1. 准备数据库

（1）根据项目要求,在 SQL Server2008 中创建 ConstructionDB 数据库、项目信息表(T_project),并向项目信息表中插入测试数据。

（2）编写 JDBC 的连接工具代码:

```
package DB;
import java.sql.*;
```

```java
public class Conn {
    public static Connection getConnection() {
        Connection conn = null;
        try {
            Class.forName("com.microsoft.sqlserver.jdbc.SQLServerDriver");
            conn = DriverManager.getConnection("jdbc:sqlserver://localhost:1433;DataBaseName=Construction-DB","sa","123456");
        } catch (Exception e) {
            e.printStackTrace();
        }
        return conn;
    }
    public static void main(String args) {
        getConnection();
    }
}
```

```java
package Dao;

import java.sql.SQLException;
import java.util.List;

import Entity.Project;

public interface ProjectDao {
    List<Project>  getAllProject() throws SQLException;
    int insertProject(Project project) throws SQLException;
    void delectProject(String id) throws SQLException;
    void updateProject(Project project) throws SQLException;
}
```

```java
package Impl;

import java.sql.Connection;
import java.sql.ResultSet;
import java.sql.SQLException;
import java.sql.Statement;
import java.util.ArrayList;
import java.util.List;

import Dao.ProjectDao;
import Entity.Project;
import DB.Conn;
```

```java
public class ProjectImpl implements ProjectDao{
private Statement state;
private Connection conn;
private ResultSet rs;
private static ProjectImpl ProjectImpl = new ProjectImpl();
    public void delectProject(String id) throws SQLException {
        String sql = "delete T_project where project_id = " + id;
        conn = Conn.getConnection();
        state = conn.createStatement();
        state.execute(sql);
    }

    public List<Project> getAllProject() throws SQLException {
      List<Project> list = new ArrayList<Project>();
      String sql = "select * from T_project";
      conn = Conn.getConnection();
      state = conn.createStatement();
      rs = state.executeQuery(sql);
      Project p = null;
      while(rs.next()){
          p = new Project();
          p.setProject_id(rs.getString("Project_id"));
          p.setProject_name(rs.getString("Project_name"));
          p.setSystem_type(rs.getString("System_type"));
          p.setInvi_dept(rs.getString("Invi_dept"));
          p.setAddr(rs.getString("Addr"));
          p.setDeputy_name(rs.getString("Deputy_name"));
          p.setTelephone(rs.getString("Telephone"));
          p.setTotal_invest(rs.getFloat("Total_invest"));
          list.add(p);
      }
      return list;
    }

    public int insertProject(Project project) throws SQLException {
        float f = project.getTotal_invest();
        String sql = "insert T_project values(" + project.getProject_id() + ",'" + project.getProject_name() + "','" + project.getInvi_dept() + "','" +
        project.getSystem_type() + "','" + project.getDeputy_name() + "','" + project.getTelephone() + "','" + project.getAddr() + "'," + project.getTotal_invest() + ")";
        if(f == 0){
            sql = "insert T_project values(" + project.getProject_id() + ",'" + project.getProject_
```

```
            name() + "','" + project.getInvi_dept() + "','" +
        project.getSystem_type() + "','" + project.getDeputy_name() + "','" + project.getTelephone()
    + "','" + project.getAddr() + "'," + null + ")";
            System.out.print(sql);
        }
        //String sql = "insert T_project values(" + project.getProject_id() + "," + project.getProject_
    name() + "," + project.getInvi_dept() + "," +
        //project.getSystem_type() + "," + project.getDeputy_name() + "," + project.getTelephone()
    + "," + project.getAddr() + "," + project.getTotal_invest() + ")";
            conn = Conn.getConnection();
            state = conn.createStatement();
            int result = state.executeUpdate(sql);
            return result;
        }


    public static ProjectImpl getInstance() {
            return ProjectImpl;
        }


    public void updateProject(Project project) throws SQLException {


        }


    }
```

2. 编写视图层代码

```
    <%@ page language = "java" contentType = "text/html; charset = GBK"
        pageEncoding = "GBK"%>
    <!DOCTYPE html PUBLIC "-//W3C//DTD HTML 4.01 Transitional//EN" "http://www.w3.
org/TR/html4/loose.dtd">
    <html>
    <head>
    <script language = "javascript">
function check() {
            var inputs = document.getElementsByTagName('input');
        for(var i = 0, len = 3; i < len; i++) {
            if(inputs[i].value.replace(/\s/g,"") == "") {
                alert('必填项不能为空！');
                inputs[i].focus();
                return false;
            }
        }
        return true;
```

```
    }
    </script>
    <meta http-equiv="Content-Type" content="text/html; charset=GBK">
    <title>Insert title here</title>

    </head>

    <body>
    <table width="834" border="0" cellpadding="0" cellspacing="0">
      <tr>
        <td height="30" background="images/title01.jpg" class="title">&gt;&gt;项目信息录入</td>
        <td width="27" height="30"><img src="images/title02.jpg" width="27" height="30"/></td>
        <td height="30" bgcolor="#029AC5" class="txt">您的位置:招投标流程 &gt; 建设工程报建 &gt; 项目信息录入</t.d>
      </tr>
</table>
<br/>
<table width="800" border="0" align="center" cellpadding="0" cellspacing="0">
  <tr>
        <td width="100%" height="30" bgcolor="#80C6FF" class="titletxt">&#8226;建设工程报建—项目信息录入(以下带<span class="txtred">*</span>为必填项)</td>
      </tr>
      <tr>
        <td height="30" align="center">
        <form id="form1" name="form1" method="post" onsubmit="return check()" action="ControlServlet?type=insert">
        <table width="100%" border="1" align="center" cellpadding="0" cellspacing="0" bgcolor="#E7E7E7">
          <tr>
            <td width="24%" height="30" align="right" class="txt"><span class="txt">工程编号:</span></td>
            <td height="30" align="left"><label for="textfield1"></label>
              <input height="20" width="400" type="text" name="textfield1" id="textfield1"/>
              <span class="txtred">*</span></td>
          </tr>
          <tr>
            <td width="24%" height="30" align="right" class="txt"><span class="txt">工程名称:</span></td>
            <td height="30" align="left"><input height="20" width="400" type="text" name="textfield2" id="textfield2"/>
```

```html
         <span class="txtred">*</span></td>
      </tr>
      <tr>
         <td height="30" align="right" class="txt"><span class="txt">报建申请单位:</span></td>
         <td height="30" align="left"><input height="20" width="400" type="text" name="textfield3" id="textfield3"/>
         <span class="txtred">*</span></td>
      </tr>
      <tr>
         <td height="30" align="right" class="txt">所有制性质:</td>
         <td height="30" align="left"><input height="20" width="200" type="text" name="textfield4" id="textfield4"/></td>
      </tr>
      <tr>
         <td height="30" align="right" class="txt">法人代表:</td>
         <td height="30" align="left"><input height="20" width="150" type="text" name="textfield5" id="textfield5"/></td>
      </tr>
      <tr>
         <td height="30" align="right" class="txt">建设单位电话:</td>
         <td height="30" align="left"><input height="20" width="200" type="text" name="textfield6" id="textfield6"/></td>
      </tr>
      <tr>
         <td height="30" align="right" class="txt">建设单位地址:</td>
         <td height="30" align="left"><span class="txtred">
            <input height="20" width="400" type="text" name="textfield7" id="textfield7"/>
            </span></td>
      </tr>
      <tr>
         <td height="30" align="right" class="txt">总投资(万元):</td>
         <td height="30" align="left"><input height="20" width="150" type="text" name="textfield8" id="textfield8"/></td>
      </tr>
   </table>
   <p><input type="submit" name="button" id="button" value="确定  "/></p>
   </form>
   <p> </p></td>
 </tr>
</table>
<p> </p>
```

```
          </body>
</html>


            <%@ page language="java" contentType="text/html;charset=gbk"
                pageEncoding="GBK"%>
            <%@ taglib prefix="c" uri="http://java.sun.com/jsp/jstl/core" %>
            <!DOCTYPE html PUBLIC "-//W3C//DTD HTML 4.01 Transitional//EN" "http://www.w3.org/TR/html4/loose.dtd">
            <html>
            <head>
            <meta http-equiv="Content-Type" content="text/html;charset=gbk">
            <title>Insert title here</title>
            </head>
            <body>
            <table width="834" border="0" cellpadding="0" cellspacing="0">
              <tr>
                <td height="30" background="images/title01.jpg" class="title">&gt;&gt;项目信息录入</td>
                <td width="27" height="30"><img src="images/title02.jpg" width="27" height="30" /></td>
                <td height="30" bgcolor="#029AC5" class="txt">您的位置:招投标流程&gt;建设工程报建&gt;项目信息录入</td>
              </tr>
            </table>
            <br />
            <table width="800" border="0" align="center" cellpadding="0" cellspacing="0">
              <tr>
                <td width="33%" height="37"> </td>
                <td width="34%" height="37"> </td>
                <td width="33%" height="37" align="right" valign="top"><a href="content_02.jsp" target="_self"><img src="images/addpro.jpg" border="0" /></a> </td>
              </tr>
              <tr>
                <td height="30" colspan="3" bgcolor="#029AC5" class="titletxt">&#8226;项目信息</td>
              </tr>
              <tr>
                <td height="30" colspan="3"><table width="100%" border="1" align="center" cellpadding="0" cellspacing="0">
                  <tr>
                    <td width="33%" height="30" align="center" bgcolor="#80C6FF"><span class="txt"><span class="titletxt">工程编号</span></span></td>
                    <td width="34%" height="30" align="center" bgcolor="#80C6FF" class="titletxt"
```

```
>工程名称</td>
                <td width="33%" height="30" align="center" bgcolor="#80C6FF" class="titletxt">相关操作</td>
            </tr>
            <c:forEach var="project" items="${list}">
            <tr>
                <td width="33%" height="30" align="center" bgcolor="#FFF5D7"><span class="txt">${project.project_id}</span></td>
                <td width="34%" height="30" align="center" bgcolor="#FFF5D7"><span class="txt">${project.project_name}</span></td>
                <td width="33%" height="30" align="center" bgcolor="#FFF5D7"><span class="txt"><a>【修改】</a><a href="ControlServlet?type=delete&&id=${project.project_id}">【删除】</a></span></td>
            </tr>
            </c:forEach>
        </table></td>
      </tr>
    </table>
    <p> </p>
  </body>
</html>


<%--
    Document: TopicList
    Created on: 2010-12-3, 8:50:30
    Author: admin
--%>

<%@page contentType="text/html" pageEncoding="GBK" import="java.util.*,Entity.*,Dao.*"%>
<%@taglib prefix="c" uri="http://java.sun.com/jsp/jstl/core"%>
<!DOCTYPE HTML PUBLIC "-//W3C//DTD HTML 4.01 Transitional//EN"
    "http://www.w3.org/TR/html4/loose.dtd">

<html>
    <head>
        <meta http-equiv="Content-Type" content="text/html; charset=GBK">
        <title>欢迎来到${board.boardName}板块!</title>
        <link href="style/style.css" rel="stylesheet" type="text/css" />
    </head>
    <body>
        <jsp:include page="Head.jsp" flush="true"/>
        <div>&gt;&gt;<a href="index.jsp">论坛首页</a>&gt;&gt;${board.boardName}
```

```
</div><br>
<div><a href="Post.jsp?boardId=${board.boardId}&boardName=${board.boardName}"><img src="image/post.gif" border="0"/></a></div><br>
<div><a href="./TopicListDisplayServlet?boardId=${board.boardId}&currentPage=${prePage}">上一页</a>|<a href="./TopicListDisplayServlet?boardId=${board.boardId}&currentPage=${nextPage}">下一页</a></div>
<div class="t">
    <table cellpadding="0" cellspacing="0" width="100%">
        <tr><th class="h" colspan="4"> </th></tr>
        <tr align="center" class="tr2">
            <td colspan="2">文章</td>
            <td width="10%">作者</td>
            <td width="10%">回复</td>
        </tr>
        <c:forEach var="topic" items="${topicList}">
        <tr align="center" class="tr3">
            <td width="5%"><img src="image/topic.gif"/></td>
            <td align="left"><a href="./ReplyListDisplayServlet?boardName=${board.boardName}&topicId=${topic.topicId}&currentPage=1">${topic.title}</a></td>
            <td>${topic.user.uName}</td>
            <td>${topic.replyCount}</td>
        </tr>
        </c:forEach>
    </table>
</div>
<div align="center"><jsp:include page="Bottom.jsp" flush="true"/></div>
</body>
</html>
```

3. 编写模型层代码

```
package Entity;

public class Project {
    private String Project_id;
    private String Project_name;
    private String Invi_dept;
    private String System_type;
    private String Deputy_name;
    private String Telephone;
    private String Addr;
    private float Total_invest;
    public String getProject_id() {
        return Project_id;
```

```java
    }
    public void setProject_id(String project_id){
        Project_id = project_id;
    }
    public String getProject_name(){
        return Project_name;
    }
    public void setProject_name(String project_name){
        Project_name = project_name;
    }
    public String getInvi_dept(){
        return Invi_dept;
    }
    public void setInvi_dept(String invi_dept){
        Invi_dept = invi_dept;
    }
    public String getSystem_type(){
        return System_type;
    }
    public void setSystem_type(String system_type){
        System_type = system_type;
    }
    public String getDeputy_name(){
        return Deputy_name;
    }
    public void setDeputy_name(String deputy_name){
        Deputy_name = deputy_name;
    }
    public String getTelephone(){
        return Telephone;
    }
    public void setTelephone(String telephone){
        Telephone = telephone;
    }
    public String getAddr(){
        return Addr;
    }
    public void setAddr(String addr){
        Addr = addr;
    }
    public float getTotal_invest(){
        return Total_invest;
    }
```

```java
    public void setTotal_invest(float total_invest) {
        Total_invest = total_invest;
    }

}
```

4. 编写控制层代码

```java
package Servlet;

import java.io.IOException;
import java.io.UnsupportedEncodingException;
import java.sql.SQLException;
import java.util.List;

import javax.servlet.ServletException;
import javax.servlet.http.HttpServletRequest;
import javax.servlet.http.HttpServletResponse;

import Dao.ProjectDao;
import Entity.Project;
import Impl.ProjectImpl;

/**
 * Servlet implementation class for Servlet: ControlServlet
 *
 */
public class ControlServlet extends javax.servlet.http.HttpServlet implements
        javax.servlet.Servlet {
    static final long serialVersionUID = 1L;

    /*
     * (non-Java-doc)
     *
     * @see javax.servlet.http.HttpServlet#HttpServlet()
     */
    public ControlServlet() {
        super();
    }

    /*
     * (non-Java-doc)
     *
     * @see javax.servlet.http.HttpServlet#doGet(HttpServletRequest request,
```

```java
 *       HttpServletResponse response)
 */
protected void doGet(HttpServletRequest request,
        HttpServletResponse response) throws ServletException, IOException {
    request.setCharacterEncoding("GBK");
    // TODO Auto-generated method stub
    start(request, response);
}

/*
 * (non-Java-doc)
 *
 * @see javax.servlet.http.HttpServlet#doPost(HttpServletRequest request,
 *       HttpServletResponse response)
 */
protected void doPost(HttpServletRequest request,
        HttpServletResponse response) throws ServletException, IOException {
    request.setCharacterEncoding("GBK");
    start(request, response);
}

private void start(HttpServletRequest request, HttpServletResponse response) throws UnsupportedEncodingException {
    String type = request.getParameter("type");
    if ("insert".equals(type)) {
        String Project_id = request.getParameter("textfield1");
        String Project_name = request.getParameter("textfield2");
        String Invi_dept = request.getParameter("textfield3");
        String System_type = request.getParameter("textfield4");
        String Deputy_name = request.getParameter("textfield5");
        String Telephone = request.getParameter("textfield6");
        String Addr = request.getParameter("textfield7");

        if (System_type.equals("") || System_type == null) {
            System_type = null;
        }
        if (Deputy_name.equals("") || Deputy_name == null) {
            Deputy_name = null;
        }
        if (Telephone.equals("") || Telephone == null) {
            Telephone = null;
        }
        if (Addr.equals("") || Addr == null) {
```

```java
                Addr = null;
            }

            float Total_invest = 0;
            try {
                Total_invest = Float.parseFloat(request
                    .getParameter("textfield8"));
            } catch (Exception ex) {
                Total_invest = 0;

            }
            Project p = new Project();
            p.setProject_id(Project_id);
            p.setAddr(Addr);
            p.setDeputy_name(Deputy_name);
            p.setInvi_dept(Invi_dept);
            p.setProject_name(Project_name);
            p.setSystem_type(System_type);
            p.setTelephone(Telephone);
            p.setTotal_invest(Total_invest);
            ProjectDao projectDao = new ProjectImpl();
            try {
                projectDao.insertProject(p);
                List<Project> list = projectDao.getAllProject();
                request.setAttribute("list", list);
request.getRequestDispatcher("project_list.jsp").forward(request, response);
            } catch (Exception e) {
                // TODO Auto-generated catch block
                e.printStackTrace();
            }

        }

        if ("delete".equals(type)) {
            String id = request.getParameter("id");
            ProjectDao projectDao = new ProjectImpl();
            try {
                projectDao.delectProject(id);
                List<Project> list = projectDao.getAllProject();
                request.setAttribute("list", list);
request.getRequestDispatcher("content_01.jsp").forward(request, response);
            } catch (Exception e) {
```

```java
            // TODO Auto-generated catch block
            e.printStackTrace();
        }
    }

    if("update".equals(type)){

    }
  }
}

package Servlet;

import java.io.IOException;
import java.sql.SQLException;
import java.util.List;

import javax.servlet.ServletException;
import javax.servlet.http.HttpServletRequest;
import javax.servlet.http.HttpServletResponse;

import Dao.ProjectDao;
import Entity.Project;
import Impl.ProjectImpl;

public class DisplayProjectsServlet extends javax.servlet.http.HttpServlet implements javax.servlet.Servlet {
    static final long serialVersionUID = 1L;

    /* (non-Java-doc)
     * @see javax.servlet.http.HttpServlet#HttpServlet()
     */
    public DisplayProjectsServlet() {
        super();
    }

    /* (non-Java-doc)
     * @see javax.servlet.http.HttpServlet#doGet(HttpServletRequest request, HttpServletResponse response)
     */
    protected void doGet(HttpServletRequest request, HttpServletResponse response) throws ServletException, IOException {
        ProjectDao projectDao = ProjectImpl.getInstance();
```

```java
        try {
            List < Project >  list = projectDao. getAllProject( ) ;
            request. setAttribute( "list" , list) ;
            request. getRequestDispatcher( "project_list. jsp" ) . forward( request, response) ;
        } catch (SQLException e) {
            // TODO Auto-generated catch block
            e. printStackTrace( ) ;
        }
    }


    /* ( non-Java-doc)
     * @ see javax. servlet. http. HttpServlet#doPost( HttpServletRequest request, HttpServletResponse response)
     */
    protected void doPost( HttpServletRequest request, HttpServletResponse response) throws ServletException, IOException {
        ProjectDao projectDao = ProjectImpl. getInstance( ) ;
        try {
            List < Project >  list = projectDao. getAllProject( ) ;
            request. setAttribute( "list" , list) ;
            request. getRequestDispatcher( "content_01. jsp" ) . forward( request, response) ;
        } catch (SQLException e) {
            // TODO Auto - generated catch block
            e. printStackTrace( ) ;
        }

    }
}
```

任务二：建设工程监管信息系统企业信息模块

一、任务描述

你作为《建设工程监管信息系统》项目开发组的程序员，请实现如下功能：
- 企业信息的列表显示；
- 企业信息的添加。

二、功能描述

(1)点击图 2.1 中的"企业信息管理"菜单项，则在右边的主体部分中显示企业信息列表。

图 2.1　企业信息列表页面

(2)点击图 2.1 中的"增加企业"按钮，则进入"企业信息录入"页面，如图 2.2 所示。
(3)对图 2.2 中"＊"号的输入部分进行必填并校验。
(4)点击图 2.2 中"确定"按钮，在企业信息表中增加一条企业信息。
(5)企业信息增加成功后，自动定位到企业信息列表页面，显示更新后的项目信息列表，如图 2.1 所示。
(6)测试程序，通过"企业信息录入"页面增加两条以上企业信息。

图 2.2　企业信息录入页面

三、要求

1. 页面实现

以提供的素材为基础,实现图 2.1、图 2.2 所示页面。

2. 数据库实现

(1)创建数据库 ConstructionDB。

(2)创建企业基本情况表(T_enterprise_info),表结构见表 2.1。

表 2.1　企业基本情况表(T_enterprise_info)表结构

字段名	字段说明	字段类型	允许为空	备注
Ent_id	企业编号	char(12)	否	主键
Ent_name	企业名称	varchar(32)	否	
Begin_date	建立时间	datetime	是	日期型
Addr	详细地址	varchar(64)	是	
Reg_capital	注册资本	decimal(12,4)	是	数值型,单位:万元

(3)在表 T_enterprise_info 插入记录,见表 2.2。

表 2.2　企业基本情况表(T_enterprise_info)记录

Ent_id	Ent_name	Begin_date	Addr	Reg_capital
200200078	长沙城建有限公司	2002－01－01	长沙市天心区	1000
200400005	天信建筑企业有限公司	2004－01－01	长沙市芙蓉区	1500

3. 功能实现

(1)功能需求如图 2.3 所示。

任务二:建设工程监管信息系统企业信息模块

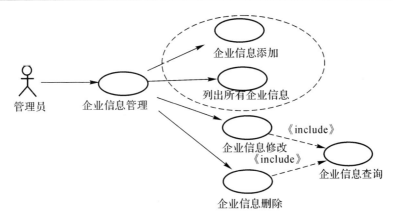

图2.3　企业信息管理模块用例图

（2）依据企业信息列表活动图完成企业信息列表显示功能,如图2.4所示。

（3）依据添加企业信息活动图完成添加企业信息功能,如图2.5所示。

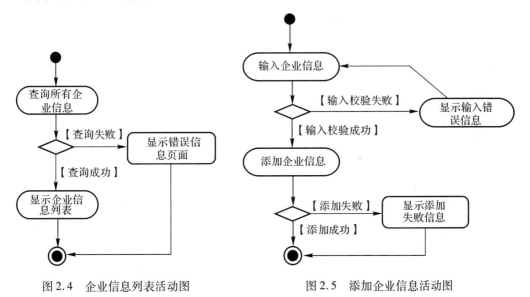

图2.4　企业信息列表活动图　　　　图2.5　添加企业信息活动图

四、必备知识

1. 数据库相关知识

（1）使用 MS SQL Server 2005/2008 创建数据库,创建数据表,设置表的字段,数据类型,主键,外键,约束。

（2）向数据表插入、删除、修改、查询数据。

2. 页面相关知识

（1）使用 HTML 制作项目页面。

（2）使用 CSS 控制页面的样式。

（3）使用 JavaScript 对页面必要的内容进行校验。

3. JSP 相关知识

（1）使用 JSTL 标准标签库控制页面显示逻辑。
（2）理解 JSP 的 request, response, session, application 的概念。
（3）使用 EL 表达式在页面显示数据。
（4）合理地使用转发和重定向控制项目的页面跳转。
（5）使用 JDBC 与数据库进行交互。
（6）MVC 模式下的分层架构，控制器，视图，模型的划分和通信。

五、解题思路

1. 数据库思路

（1）根据项目要求创建数据库和数据表，向数据表中插入合适的测试数据。
（2）导入 JDBC 驱动包，编写 JDBC 的连接工具代码。
（3）编写数据操作对象代码，负责进行与数据库的交互操作。

2. 视图层思路

（1）将提供的素材页面改写为 JSP 页面。
（2）JSP 使用 JSTL 和 EL 负责控制页面显示的逻辑。

3. 控制层思路

（1）使用 Servlet 类控制一次请求响应过程的处理。
（2）由 Servlet 按照顺序进行请求的处理、数据库交互、模型存取和封装、页面跳转逻辑控制等。

4. 模型层思路

使用 JavaBean 作为模型层，封装数据和行为。

六、操作步骤

1. 准备数据库

（1）根据项目要求，在 SQL Server2008 中创建 ConstructionDB 数据库、企业基本情况表（T_enterprise_info），并向企业基本情况表中插入测试数据。
（2）编写 JDBC 的连接工具代码。

```
package construct.dao;

import java.sql.*;
import java.io.*;
import java.util.*;
import javax.naming.Context;
import javax.naming.InitialContext;
import javax.naming.NamingException;
```

```java
import javax.sql.DataSource;

/**
 * 数据库访问操作类
 * @author
 */
public abstract class BaseDao {

    protected PreparedStatement pstmt = null; //SQL
    protected ResultSet rs = null;//结果集
    protected Connection conn = null;//数据库连接

    /**
     * 创建数据库连接
     * @return
     */
    public Connection getConn() throws SQLException, ClassNotFoundException, IOException {
        Connection conn = null;
        String dbDriver = "";
        String dbUrl = "";
        String dbUser = "sa";
        String dbPass = "sasa";
        InputStream in = this.getClass().getClassLoader().getResourceAsStream("construct/dao/construct.properties");
        if (null != in) {
            Properties props = new Properties();
            props.load(in);
            dbDriver = props.getProperty("construct.driver");//取数据库驱动程序
            dbUrl = props.getProperty("construct.url");
            dbUser = props.getProperty("construct.user.name");
            dbPass = props.getProperty("construct.user.password");
            Class.forName(dbDriver);
            conn = DriverManager.getConnection(dbUrl, dbUser, dbPass);
        }
        return conn;
    }

    /**
     * 关闭数据表操作
     */
    public void closeAll() {
        try {
            if (rs != null) {
```

```java
                rs.close();
            }
            if (pstmt != null) {
                pstmt.close();
            }
            if (conn != null) {
                conn.close();
            }
        } catch (SQLException er) {
            er.printStackTrace();
        }
    }

    /**
     * 执行SQL语句方法
     * @param preparedSql
     * @param params
     * @return
     */
    public int executeSQL(String preparedSql, String[] params) {
        int result = 0;
        this.pstmt = null;
        try {
            conn = this.getConn();
            pstmt = conn.prepareStatement(preparedSql);
            int i = 1;//计数器
            if (params != null) {//判断参数集是否为空
                for (String param : params) {//添加SQL语句参数
                    pstmt.setString(i, param);
                    i++;
                }
            }
            result = pstmt.executeUpdate();//执行SQL语句
        } catch (Exception er) {
            er.printStackTrace();
        } finally {
            this.closeAll();
        }
        return result;
    }

    /**
     * 查询语句执行方法
```

```
 * @param preparedSql SQL 语句
 * @param params 参数
 * @return
 */
public List executeQuery(String preparedSql, String[] params) {
    List results = null;
    pstmt = null;
    try {
        conn = this.getConn();
        pstmt = conn.prepareStatement(preparedSql);
        int i = 1;//计数器
        if (params != null) {//判断参数集是否为空
            for (String param : params) {//添加 SQL 语句参数
                pstmt.setString(i, param);
                i++;
            }
        }
        rs = pstmt.executeQuery();
        results = createObject(rs);//提取查询结果集
    } catch (Exception er) {
        er.printStackTrace();
    } finally {
        this.closeAll();
    }
    return results;
}

/**
 * 根据查询条件统计记录数
 * @param preparedSql
 * @param params
 * @return
 */
public int count(String preparedSql, String[] params) {
    int result = 0;
    pstmt = null;
    try {
        conn = this.getConn();
        pstmt = conn.prepareStatement(preparedSql);
        int i = 1;//计数器
        if (params != null) {//判断参数集是否为空
            for (String param : params) {//添加 SQL 语句参数
                pstmt.setString(i, param);
```

```java
                    i++;
                }
            }
            rs = pstmt.executeQuery();//执行统计SQL语句
            if(rs.next()){//获取统计结果
                result = rs.getInt(1);
            }
        }catch(Exception er){
            er.printStackTrace();
        }finally{
            this.closeAll();
        }
        return result;
    }

    /**
     *从结果集中组装对象方法
     *@param rs 查询结果集
     *@return
     */
    public abstract List createObject(ResultSet rs) throws SQLException;
}


/*
 * To change this template, choose Tools | Templates
 * and open the template in the editor.
 */

package construct.dao;

import construct.entity.Enterprise;
import java.sql.ResultSet;
import java.sql.SQLException;
import java.util.ArrayList;
import java.util.List;

/**
 *
 * @author Administrator
 */
public class EnterpriseDao extends BaseDao{

    @Override
```

```java
public List createObject(ResultSet rs) throws SQLException {
    List<Enterprise> enterpiseList = new ArrayList<Enterprise>();
    while(rs.next())
    {
        Enterprise enterprise = new Enterprise();
        enterprise.setEntId(rs.getString("Ent_id"));
        enterprise.setEndName(rs.getString("Ent_name"));
        enterprise.setBeginDate(rs.getDate("Begin_date").toString());
        enterprise.setRegCapital(rs.getFloat("Reg_capital"));
        enterpiseList.add(enterprise);
    }
    return enterpiseList;
}

public int addEnterprise(Enterprise enterprise) {
    String addStr = "insert into T_enterprise_info (ent_id,Ent_name,Begin_date,Addr,Reg_capital) values (?,?,?,?,?)";
    String[] params = {enterprise.getEntId(),enterprise.getEndName(),enterprise.getBeginDate(),enterprise.getAddr(),String.valueOf(enterprise.getRegCapital())};
    return this.executeSQL(addStr, params);
}

public List findListEnterprise() {
    String strQuery = "select * from T_enterprise_info";
    return this.executeQuery(strQuery, null);
}
}
```

```
# To change this template, choose Tools | Templates
# and open the template in the editor.
construct.driver = com.microsoft.sqlserver.jdbc.SQLServerDriver
construct.url = jdbc:sqlserver://localhost:1433;DataBaseName = ConstructionDB
construct.user.name = sa
construct.user.password = 123456
```

2. 编写视图层代码

```
<%@ page contentType="text/html" pageEncoding="UTF-8" import="java.util.*,construct.entity.Enterprise"%>
<!DOCTYPE HTML PUBLIC "-//W3C//DTD HTML 4.01 Transitional//EN"
    "http://www.w3.org/TR/html4/loose.dtd">
<html xmlns="http://www.w3.org/1999/xhtml">
<head>
```

```html
<meta http-equiv="Content-Type" content="text/html; charset=utf-8" />
<title>无标题文档</title>
<LINK href="css/content.css" type=text/css rel=stylesheet>
</head>

<body>
<table width="834" border="0" cellpadding="0" cellspacing="0">
  <tr>
    <td height="30" background="images/title01.jpg" class="title">&gt;&gt;企业信息列表</td>
    <td width="27" height="30"><img src="./images/title02.jpg" width="27" height="30" /></td>
    <td height="30" bgcolor="#029AC5" class="txt"><p>您的位置:业务管理 &gt;企业信息管理</p>
    </td>
  </tr>
</table>
<br />
<table width="800" border="0" align="center" cellpadding="0" cellspacing="0">
  <tr>
    <td width="33%" height="37"> </td>
    <td width="34%" height="37"> </td>
    <td width="33%" height="37" align="right" valign="top">
      <a href="enterpriseAdd.jsp" target="_self"><img src="images/addEnter.JPG" border="0" /></a> </td>
  </tr>
  <tr>
    <td height="30" colspan="3" bgcolor="#029AC5" class="titletxt">&#8226;企业信息</td>
  </tr>
  <tr>
    <td height="30" colspan="3"><table width="100%" border="1" align="center" cellpadding="0" cellspacing="0">
      <tr>
        <td width="33%" height="30" align="center" bgcolor="#80C6FF"><span class="txt"><span class="titletxt">企业编号</span></span></td>
        <td width="34%" height="30" align="center" bgcolor="#80C6FF" class="titletxt">企业名称</td>
        <td width="33%" height="30" align="center" bgcolor="#80C6FF" class="titletxt">相关操作</td>
      </tr>
      <%
```

```jsp
                List<Enterprise> entList=(List<Enterprise>)request.getAttribute("entList");
                if(entList!=null)
                    {
                for(Enterprise ent:entList)
                    {
        %>
                <tr>
                    <td width="33%" height="30" align="center" bgcolor="#FFF5D7"><span class="txt"><%=ent.getEntId()%></span></td>
                    <td width="34%" height="30" align="center" bgcolor="#FFF5D7"><span class="txt"><%=ent.getEndName()%></span></td>
                    <td width="33%" height="30" align="center" bgcolor="#FFF5D7"><span class="txt">【修改】【删除】</span></td>
                </tr>
        <%
            }}
        %>

    </table></td>
    </tr>
</table>
<p> </p>
</body>
</html>


    <%@ page contentType="text/html" pageEncoding="UTF-8"%>
    <!DOCTYPE HTML PUBLIC "-//W3C//DTD HTML 4.01 Transitional//EN"
        "http://www.w3.org/TR/html4/loose.dtd">
    <html xmlns="http://www.w3.org/1999/xhtml">
    <head>
    <meta http-equiv="Content-Type" content="text/html; charset=utf-8" />
    <title>无标题文档</title>
    <LINK
href="css/content.css" type=text/css rel=stylesheet>
    <script language="javascript">
        function check()
        {
            if(document.form1.entId.value=="")
                {
                    alert("企业编号不能为空");
                    return false;
                }
            if(document.form1.entName.value=="")
```

```html
                }
                    alert("企业名称不能为空");
                    return false;
                }
                    return true;
            }
        </script>
    </head>

    <body>
        <table width="834" border="0" cellpadding="0" cellspacing="0">
            <tr>
                <td height="30" background="images/title01.jpg" class="title">&gt;&gt;企业信息录入</td>
                <td width="27" height="30"><img src="images/title02.jpg" width="27" height="30" /></td>
                <td height="30" bgcolor="#029AC5" class="txt">您的位置:业务管理 &gt;企业信息管理</td>
            </tr>
        </table>
        <br />
        <table width="800" border="0" align="center" cellpadding="0" cellspacing="0">
            <tr>
                <td width="100%" height="30" bgcolor="#80C6FF" class="titletxt">&#8226;企业信息管理—企业信息录入(以下带<span class="txtred">*</span>为必填项)</td>
            </tr>
            <tr>
                <td height="30" align="center">
                    <form id="form1" name="form1" method="post" action="EnterpriseServlet?type=add" onsubmit="return check()">
                        <table width="100%" border="1" align="center" cellpadding="0" cellspacing="0" bgcolor="#E7E7E7">
                            <tr>
                                <td width="24%" height="30" align="right" class="txt"><span class="txt">企业编号:</span></td>
                                <td height="30" align="left"><label for="textfield"></label>
                                    <input name="entId" type="text" id="entId" size="40" width="400" height="20" />
                                    <span class="txtred">*</span></td>
                            </tr>
                            <tr>
                                <td width="24%" height="30" align="right" class="txt"><span class="txt">企业名称:</span></td>
```

```html
            <td height="30" align="left"><input name="entName" type="text" id="entName" size="40" width="400" height="20" />
                <span class="txtred">*</span></td>
        </tr>

        <tr>
            <td height="30" align="right" class="txt">建立时间：</td>
            <td height="30" align="left"><input height="20" width="200" type="text" name="beginDate" id="beginDate" /></td>
        </tr>
        <tr>
            <td height="30" align="right" class="txt">详细地址：</td>
            <td height="30" align="left"><input name="addr" type="text" id="addr" size="40" height="20" /></td>
        </tr>
        <tr>
            <td height="30" align="right" class="txt">注册资本(万元)：</td>
            <td height="30" align="left"><span class="txtred">
                <input height="20" width="100" type="text" name="regCapital" id="regCapital" />
            </span></td>
        </tr>

    </table>
    <p><input type="submit" name="button" id="button" value="确  定" /></p>
    </form>
    <p> </p></td>
    </tr>
    </table>
    <p> </p>
    </body>
</html>
```

3. 编写模型层代码

```
/*
 * To change this template, choose Tools | Templates
 * and open the template in the editor.
 */

package construct.entity;
```

```java
/**
 * 
 * @author Administrator
 */
public class Enterprise {
    private String entId;
    private String endName;
    private String beginDate;
    private String addr;
    private float regCapital;

    public String getAddr() {
        return addr;
    }

    public void setAddr(String addr) {
        this.addr = addr;
    }

    public String getBeginDate() {
        return beginDate;
    }

    public void setBeginDate(String beginDate) {
        this.beginDate = beginDate;
    }

    public String getEndName() {
        return endName;
    }

    public void setEndName(String endName) {
        this.endName = endName;
    }

    public String getEntId() {
        return entId;
    }

    public void setEntId(String entId) {
        this.entId = entId;
    }
```

```java
    public float getRegCapital() {
        return regCapital;
    }

    public void setRegCapital(float regCapital) {
        this.regCapital = regCapital;
    }

}
```

4. 编写控制层代码

```java
/*
 * To change this template, choose Tools | Templates
 * and open the template in the editor.
 */

package construct.servlet;

import construct.dao.EnterpriseDao;
import construct.entity.Enterprise;
import java.io.IOException;
import java.io.PrintWriter;
import java.util.List;
import javax.servlet.ServletException;
import javax.servlet.annotation.WebServlet;
import javax.servlet.http.HttpServlet;
import javax.servlet.http.HttpServletRequest;
import javax.servlet.http.HttpServletResponse;

/**
 *
 * @author Administrator
 */
@WebServlet(name = "EnterpsieServlet", urlPatterns = {"/EnterpriseServlet"})
public class EnterpriseServlet extends HttpServlet {

    /**
     * Processes requests for both HTTP <code>GET</code> and <code>POST</code> methods.
     * @param request servlet request
     * @param response servlet response
     * @throws ServletException if a servlet-specific error occurs
     * @throws IOException if an I/O error occurs
```

```java
 */
protected void processRequest(HttpServletRequest request, HttpServletResponse response)
throws ServletException, IOException {
    response.setContentType("text/html;charset=UTF-8");
    String type = request.getParameter("type");
    EnterpriseDao enterpriseDao = new EnterpriseDao();
    if("list".equals(type))
    {
        List<Enterprise> entList = enterpriseDao.findListEnterprise();
        request.setAttribute("entList", entList);
        request.getRequestDispatcher("enterpriseList.jsp").forward(request, response);
    }
    if("add".equals(type))
    {
        String entId = request.getParameter("entId");
        String entName = request.getParameter("entName");
        String beginDate = request.getParameter("beginDate");
        String addr = request.getParameter("addr");
        String regCapital = request.getParameter("regCapital");
        Enterprise ent = new Enterprise();
        ent.setEndName(entName);
        ent.setEntId(entId);
        ent.setBeginDate(beginDate);
        ent.setAddr(addr);
        ent.setRegCapital(Float.valueOf(regCapital));
        enterpriseDao.addEnterprise(ent);

request.getRequestDispatcher("/EnterpriseServlet?type=list").forward(request, response);

    }
}


// <editor-fold defaultstate="collapsed" desc="HttpServlet methods. Click on the + sign on the left to edit the code.">
/**
 * Handles the HTTP <code>GET</code> method.
 * @param request servlet request
 * @param response servlet response
 * @throws ServletException if a servlet-specific error occurs
 * @throws IOException if an I/O error occurs
 */
@Override
protected void doGet(HttpServletRequest request, HttpServletResponse response)
```

```java
        throws ServletException, IOException {
    processRequest(request, response);
}


/**
 * Handles the HTTP <code>POST</code> method.
 * @param request servlet request
 * @param response servlet response
 * @throws ServletException if a servlet-specific error occurs
 * @throws IOException if an I/O error occurs
 */
@Override
protected void doPost(HttpServletRequest request, HttpServletResponse response)
        throws ServletException, IOException {
    processRequest(request, response);
}


/**
 * Returns a short description of the servlet.
 * @return a String containing servlet description
 */
@Override
public String getServletInfo() {
    return "Short description";
}//


}
```

任务三：建设工程监管信息系统信息列表模块

一、任务描述

你作为《建设工程监管信息系统》项目开发组的程序员，请实现如下功能：
➢ 招标项目信息列表显示；
➢ 评委信息列表显示。

二、功能描述

（1）点击图 3.1 中的"抽取评委"菜单项，则在右边的主体部分中显示所有的招标项目信息及相关操作。

图 3.1 招标项目信息列表页面

（2）点击图 3.1 中的某项目的"已选评委"链接，进入该项目的"已选评委列表"页面，如图 3.2 所示。

（3）在图 3.2 的已选评委列表中显示所选项目的所有评委信息。

三、要求

1.页面实现

以提供的素材为基础，实现图 3.1、图 3.2 所示页面。

任务三:建设工程监管信息系统信息列表模块

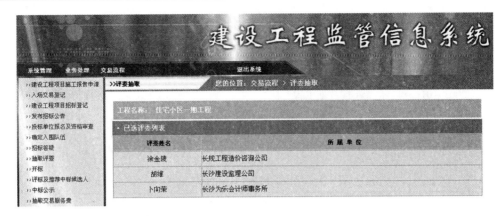

图 3.2　已选评委列表页面

2. 数据库实现

(1)创建数据库 ConstructionDB。

(2)创建招标项目信息表(T_project),表结构见表 3.1。

表 3.1　招标项目信息表(T_project)表结构

字段名	字段说明	字段类型	允许为空	备注
Project_id	工程编号	varchar(32)	否	主键
Project_name	工程名称	varchar(64)	否	
Deputy_name	法人代表	varchar(16)	是	
Telephone	电话	varchar(16)	是	
Addr	地址	varchar(64)	是	

(3)创建评委信息表(T_expert),表结构见表 3.2。

表 3.2　评委信息表(T_expert)表结构

字段名	字段说明	字段类型	是否允许为空	备注
Expert_id	专家 ID	char(5)	否	主键
Expert_name	专家姓名	char(12)	否	
Id_card	身份证	char(18)	否	
Dept	单位	varchar(64)	是	

(4)创建招标项目-评委关联信息表(T_project_expert),表结构见表 3.3。

表 3.3　招标项目-评委关联信息表(T_project_expert)表结构

字段名	字段说明	字段类型	是否允许为空	备注
ID	序号	int	否	主键,从1开始自增
Project_id	工程编号	varchar(32)	否	外键
Expert_id	专家 ID	char(5)	否	外键

(5)在表 T_project 插入记录,见表 3.4。

表 3.4 招标项目信息表(T_project)记录

Project_id	Project_name	Deputy_name	Telephone	Addr
2003-01	住宅小区一期工程	张三	2626266	长沙市天心区
2003-02	教学大楼	王平	8374777	长沙市芙蓉区

(6)在表 T_expert 插入记录,见表 3.5。

表 3.5 评委信息表(T_expert)记录

Expert_id	Expert_name	Id_card	Sex	Dept
E1001	涂金陵	430911196901228740	男	长规工程造价咨询公司
E1002	胡维	430923197011228000	男	长沙建设监理公司
E1003	刘莉莉	430234196907224560	女	长沙交通学院
E1004	卜向荣	430345197801234530	女	长沙为乐会计师事务所

(7)在表 T_project_expert 插入记录,见表 3.6。

表 3.6 招标项目-评委关联信息表(T_project_expert)记录

ID	Project_id	Expert_id
1	2003-01	E1001
2	2003-01	E1002
3	2003-01	E1004
4	2003-02	E1004

3. 功能实现

(1)功能需求如图 3.3 所示。

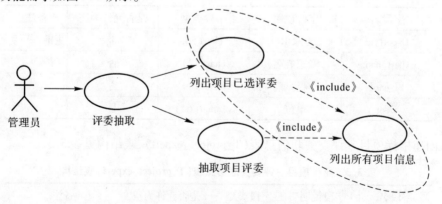

图 3.3 评委抽取模块用例图

(2)依据项目信息列表活动图,完成项目信息列表显示功能,如图 3.4 所示。
(3)依据评委信息列表活动图,完成评委信息列表显示功能,如图 3.5 所示。

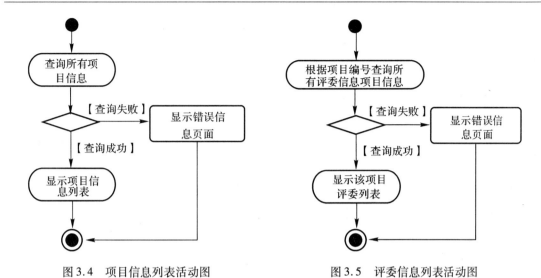

图 3.4　项目信息列表活动图　　　　图 3.5　评委信息列表活动图

四、必备知识

1. 数据库相关知识

（1）使用 MS SQL Server 2005/2008 创建数据库,创建数据表,设置表的字段,数据类型,主键,外键,约束。

（2）向数据表插入、删除、修改、查询数据。

2. 页面相关知识

（1）使用 HTML 制作项目页面。

（2）使用 CSS 控制页面的样式。

（3）使用 JavaScript 对页面必要的内容进行校验。

3. JSP 相关知识

（1）使用 JSTL 标准标签库控制页面显示逻辑。

（2）理解 JSP 的 request,response,session,application 的概念。

（3）使用 EL 表达式在页面显示数据。

（4）合理地使用转发和重定向控制项目的页面跳转。

（5）使用 JDBC 与数据库进行交互。

（6）MVC 模式下的分层架构,控制器,视图,模型的划分和通信。

五、解题思路

1. 数据库思路

（1）根据项目要求创建数据库和数据表,向数据表中插入合适的测试数据。

（2）导入 JDBC 驱动包,编写 JDBC 的连接工具代码。

（3）编写数据操作对象代码,负责进行与数据库的交互操作。

2. 视图层思路

（1）将提供的素材页面改写为 JSP 页面。

（2）JSP 使用 JSTL 和 EL 负责控制页面显示的逻辑。

3. 控制层思路

（1）使用 Servlet 类控制一次请求响应过程的处理。

（2）由 Servlet 按照顺序进行请求的处理、数据库交互、模型存取和封装、页面跳转逻辑控制等。

4. 模型层思路

使用 JavaBean 作为模型层，封装数据和行为。

六、操作步骤

1. 准备数据库

（1）根据项目要求，在 SQL Server2008 中创建 ConstructionDB 数据库、招标项目信息表(T_project)，评委信息表(T_expert)，招标项目－评委关联信息表(T_project_expert)，并插入测试数据。

（2）编写 JDBC 的连接工具代码。

```
/*
 * To change this template, choose Tools | Templates
 * and open the template in the editor.
 */

package util;

import java.io.InputStream;
import java.sql.Connection;
import java.sql.DriverManager;
import java.util.Properties;

/**
 *
 * @author zz
 */
public class DBUtil
{
    /**
     * 一个对数据源的连接
     */
    private Connection con;
```

```java
        private String DRIVER;
        private String URL;
        private String DBNAME;
        private   String DBPASS;
    public static int pagesize = 10;

        public DBUtil() throws Exception
        {
                InputStream in = this.getClass().getClassLoader().getResourceAsStream("util/bbs.properties");
                if(null! = in)
                {
                    Properties props = new Properties();
                    //从属性文件中读取属性
                    props.load(in);
                    DRIVER = props.getProperty("bbs.driver");
                    URL = props.getProperty("bbs.url");
                    DBNAME = props.getProperty("bbs.user.name");
                    DBPASS = props.getProperty("bbs.user.password");
                }

        }

        /**
         * (getConnecetion) <BR>
         * 取数据库连接 <BR>
         * @return Connection －数据库连接 <BR>
         */
        public Connection getConnecetion() throws Exception
        {
            Connection conn = null;
            try
            {
                if(this.con! = null&&! con.isClosed())
                {
                    conn = con;
                }
                else
                {
                    System.out.println(DRIVER);
                    //加载驱动程序
```

```
                    Class.forName(DRIVER);

                    //创建连接
                    conn = DriverManager.getConnection(URL, DBNAME, DBPASS);

                }

            }
        catch(Exception ex)
        {
                System.out.println(ex);
                throw ex;
        }
        return conn;
    }


    /**
     *(closeConnection) <BR>
     *关闭数据库连接 <BR>
     */
    public void closeConnection() throws Exception
    {

        try
        {
            if(this.con ! = null)
            {
                if(this.con.isClosed() = = false)
                {
                    this.con.close();
                }
            }

        }
        catch(Exception e)
        {
            System.out.println("释放连接时出错:" + e);
            throw e;
        }
    }
```

}

2. 编写视图层代码

```jsp
<%--
    Document   : project_list
    Created on : 2009-6-10, 15:17:54
    Author     : admin
--%>

<%@ page contentType="text/html" pageEncoding="UTF-8"%>
<!DOCTYPE HTML PUBLIC "-//W3C//DTD HTML 4.01 Transitional//EN"
    "http://www.w3.org/TR/html4/loose.dtd">

<html xmlns="http://www.w3.org/1999/xhtml">
<head>
<meta http-equiv="Content-Type" content="text/html; charset=utf-8" />
<title>无标题文档</title>
<LINK href="../css/content.css" type="text/css" rel="stylesheet">
<%@ page import="java.util.ArrayList" %>
<%@ page import="Bean.*" %>
</head>

<body>
<table width="834" border="0" cellpadding="0" cellspacing="0">
  <tr>
    <td height="30" background="../images/title01.jpg" class="title">&gt;&gt;项目信息</td>
    <td width="27" height="30"><img src="images/title02.jpg" width="27" height="30" /></td>
    <td height="30" bgcolor="#029AC5" class="txt"><span class="titletxt">您的位置:交易流程&gt;评委抽取</span></td>
  </tr>
</table>
<br />
<table width="800" border="0" align="center" cellpadding="0" cellspacing="0">
  <tr>
    <td width="33%" height="37"> </td>
    <td width="34%" height="37"> </td>
    <td width="33%" height="37" align="right" valign="top"><a href="content_02.html" target="_self"></a> </td>
  </tr>
  <tr>
```

```
            <td height="30" colspan="3" bgcolor="#029AC5" class="titletxt">&#8226;项目信息</td>
          </tr>
          <tr>
            <td width="33%" height="30" align="center" bgcolor="#80C6FF"><span class="txt"><span class="titletxt">工程编号</span></span></td>
            <td width="34%" height="30" align="center" bgcolor="#80C6FF" class="titletxt">工程名称</td>
            <td width="33%" height="30" align="center" bgcolor="#80C6FF" class="titletxt">相关操作</td>
          </tr>
          <tr>
            <td height="30" colspan="3">
              <table width="100%" border="1" align="center" cellpadding="0" cellspacing="0">
                <%
                    ArrayList<project> projects=(ArrayList<project>)request.getAttribute("msg");
                    for(project pro:projects){
                %>
                <tr>
                  <td width="33%" height="30" align="center" bgcolor="#FFFFFF"><span class="txt"><%=pro.getProject_id()%></span></td>
                  <td width="34%" height="30" align="center" bgcolor="#FFFFFF"><span class="txt"><%=pro.getProject_name()%></span></td>
                  <td width="33%" height="30" align="center" bgcolor="#FFFFFF"><a href="../Exam4-3/expert_select?proId=<%=pro.getProject_id()%>" target="_self">【已选评委】</a> 【评委抽取】</td>
                </tr>
                <%
                    }
                %>
              </table>
            </td>
          </tr>
        </table>
        <p> </p>
    </body>
</html>

<%--
```

```
    Document    : expert_select
    Created on  : 2009-6-10, 16:04:11
    Author      : admin
--%>

<%@ page contentType="text/html" pageEncoding="UTF-8"%>
<!DOCTYPE HTML PUBLIC "-//W3C//DTD HTML 4.01 Transitional//EN"
    "http://www.w3.org/TR/html4/loose.dtd">

<html xmlns="http://www.w3.org/1999/xhtml">
<head>
<meta http-equiv="Content-Type" content="text/html; charset=utf-8" />
<title>无标题文档</title>
<LINK href="../css/content.css" type="text/css" rel="stylesheet"/>
<%@ page import="java.util.*"%>
<%@ page import="Bean.*"%>
</head>

<body>
<table width="834" border="0" cellpadding="0" cellspacing="0">
  <tr>
    <td height="30" background="images/title01.jpg" class="title"><span class="title">&gt;&gt;评委抽取</span></td>
    <td width="27" height="30"><img src="images/title02.jpg" width="27" height="30"/></td>
    <td height="30" bgcolor="#029AC5" class="txt"><span class="title"><span class="titletxt">您的位置:交易流程&gt;评委抽取</span></span></td>
  </tr>
</table>
<br/>
<table width="790" border="0" align="center" cellpadding="0" cellspacing="0">
  <tr>
    <td width="91" height="38" valign="middle" bgcolor="#80C6FF" class="titletxt" align="center">工程名称:</td>
    <td width="703" valign="middle" bgcolor="#80C6FF" class="titletxt">住宅小区一期工程</td>
  </tr>
</table>

<table width="800" border="0" align="center">
  <tr>
    <td width="40%" valign="top"><table width="100%" border="0" align="center"
```

```
cellpadding = "0" cellspacing = "0" >
    < tr >
        < td width = "100%" height = "30" bgcolor = "#029AC5" class = "titletxt" >  &#8226;已选评委列表 </td>
    </tr>
    < tr >
        < td height = "30" align = "center" > < form id = "form2" name = "form1" method = "post" action = "project_expert_list.html" >
            < table width = "100%" border = "1" align = "center" cellpadding = "0" cellspacing = "0" bgcolor = "#E7E7E7" >
                < tr >
                    < td width = "16%" height = "30" align = "center" bgcolor = "#80C6FF" class = "txt" > < span class = "title" >评委姓名 </span> </td>
                    < td width = "50%" align = "center" bgcolor = "#80C6FF" class = "title" > < span class = "txt" > < span class = "title" >所属单位 </span> </span> </td>
                </tr>

                < %
                ArrayList < expert > experts = ( ArrayList < expert > ) request.getAttribute( "msg" ) ;
                for( expert exp : experts ) {

                % >
                < tr >
                    < td width = "16%" height = "30" align = "center" class = "txt" > < % = exp.getExpert_name( ) % > </td>
                    < td align = "left" class = "txt" > < % = exp.getDept( ) % > </td>
                </tr>
                < %
                }
                % >

            </table>
        </form>
        < p >   </p> </td>
    </tr>
</table> </td>
</tr>
</table>
< p >   </p>
</body>
</html>
```

3. 编写模型层代码

```java
/*
 * To change this template, choose Tools | Templates
 * and open the template in the editor.
 */

package Bean;

/**
 *
 * @author admin
 */
public class project {
    private String Project_id;
    private String Project_name;
    private String Deputy_name;
    private String Telephone;
    private String Addr;

    /**
     * @return the Project_id
     */
    public String getProject_id() {
        return Project_id;
    }

    /**
     * @param Project_id the Project_id to set
     */
    public void setProject_id(String Project_id) {
        this.Project_id = Project_id;
    }

    /**
     * @return the Project_name
     */
    public String getProject_name() {
        return Project_name;
    }

    /**
     * @param Project_name the Project_name to set
```

```java
     */
    public void setProject_name(String Project_name) {
        this.Project_name = Project_name;
    }

    /**
     * @return the Deputy_name
     */
    public String getDeputy_name() {
        return Deputy_name;
    }

    /**
     * @param Deputy_name the Deputy_name to set
     */
    public void setDeputy_name(String Deputy_name) {
        this.Deputy_name = Deputy_name;
    }

    /**
     * @return the Telephone
     */
    public String getTelephone() {
        return Telephone;
    }

    /**
     * @param Telephone the Telephone to set
     */
    public void setTelephone(String Telephone) {
        this.Telephone = Telephone;
    }

    /**
     * @return the Addr
     */
    public String getAddr() {
        return Addr;
    }

    /**
     * @param Addr the Addr to set
     */
```

```java
    public void setAddr(String Addr){
        this.Addr = Addr;
    }

}

/*
 * To change this template, choose Tools | Templates
 * and open the template in the editor.
 */

package Bean;

/**
 *
 * @author admin
 */
public class expert{
private String Expert_id;
    private String Expert_name;
    private String ID_Card;
    private String Sex;
    private String Dept;

    /**
     * @return the Expert_id
     */
    public String getExpert_id(){
        return Expert_id;
    }

    /**
     * @param Expert_id the Expert_id to set
     */
    public void setExpert_id(String Expert_id){
        this.Expert_id = Expert_id;
    }

    /**
     * @return the Expert_name
     */
    public String getExpert_name(){
        return Expert_name;
```

```
    }

    /**
     * @param Expert_name the Expert_name to set
     */
    public void setExpert_name(String Expert_name) {
        this.Expert_name = Expert_name;
    }

    /**
     * @return the ID_Card
     */
    public String getID_Card() {
        return ID_Card;
    }

    /**
     * @param ID_Card the ID_Card to set
     */
    public void setID_Card(String ID_Card) {
        this.ID_Card = ID_Card;
    }

    /**
     * @return the Sex
     */
    public String getSex() {
        return Sex;
    }

    /**
     * @param Sex the Sex to set
     */
    public void setSex(String Sex) {
        this.Sex = Sex;
    }

    /**
     * @return the Dept
     */
    public String getDept() {
        return Dept;
    }
```

```
    /**
     * @param Dept the Dept to set
     */
    public void setDept(String Dept) {
        this.Dept = Dept;
    }
}
```

4. 编写控制层代码

```
/*
 * To change this template, choose Tools | Templates
 * and open the template in the editor.
 */

package Servlet;

import Bean.project;
import java.io.IOException;
import java.io.PrintWriter;
import java.sql.Connection;
import java.sql.ResultSet;
import java.sql.Statement;
import java.util.ArrayList;
import javax.servlet.RequestDispatcher;
import javax.servlet.ServletException;
import javax.servlet.annotation.WebServlet;
import javax.servlet.http.HttpServlet;
import javax.servlet.http.HttpServletRequest;
import javax.servlet.http.HttpServletResponse;
import util.DBUtil;

/**
 *
 * @author admin
 */
@WebServlet(name = "project_list", urlPatterns = {"/project_list"})
public class project_list extends HttpServlet {

    /**
     * Processes requests for both HTTP <code>GET</code> and <code>POST</code> methods.
     * @param request servlet request
```

```
 * @param response servlet response
 * @throws ServletException if a servlet-specific error occurs
 * @throws IOException if an I/O error occurs
 */
protected void processRequest(HttpServletRequest request, HttpServletResponse response)
        throws ServletException, IOException {
    response.setContentType("text/html;charset=UTF-8");
    PrintWriter out = response.getWriter();
    try {
        /* TODO output your page here
        out.println("<html>");
        out.println("<head>");
        out.println("<title>Servlet project_list</title>");
        out.println("</head>");
        out.println("<body>");
        out.println("<h1>Servlet project_list at " + request.getContextPath() + "</h1>");
        out.println("</body>");
        out.println("</html>");
        */
    } finally {
        out.close();
    }
}

// <editor-fold defaultstate="collapsed" desc="HttpServlet methods. Click on the + sign on the left to edit the code.">
/**
 * Handles the HTTP <code>GET</code> method.
 * @param request servlet request
 * @param response servlet response
 * @throws ServletException if a servlet-specific error occurs
 * @throws IOException if an I/O error occurs
 */
@Override
protected void doGet(HttpServletRequest request, HttpServletResponse response)
        throws ServletException, IOException {
    try {

        boolean isLogin = false;
        //以下去访问数据库

        //获得数据库连接
```

```java
            DBUtil dbUtil = new DBUtil();
            //创建连接
            Connection con = dbUtil.getConnecetion();

            Statement stat = con.createStatement();
            //查询表
            ResultSet rs = stat.executeQuery("select * from T_project");

            String strInfo = null;
            ArrayList<project> projects = new ArrayList<project>();
            while(rs.next()){
            project pro = new project();
            pro.setProject_id(rs.getString("Project_id"));
            pro.setProject_name(rs.getString("Project_name"));
            pro.setDeputy_name(rs.getString("Deputy_name"));
            pro.setTelephone(rs.getString("Telephone"));
            pro.setAddr(rs.getString("Addr"));
            projects.add(pro);
            }
            //关闭数据库的连接
            dbUtil.closeConnection();
            request.setAttribute("msg",projects);
                RequestDispatcher rd =
                    request.getRequestDispatcher("/jsp/project_list.jsp");
                rd.forward(request, response);

                return;
        } catch (Exception e) {
            System.out.println(e);
        }
    }
}

/**
 * Handles the HTTP <code>POST</code> method.
 * @param request servlet request
 * @param response servlet response
 * @throws ServletException if a servlet-specific error occurs
 * @throws IOException if an I/O error occurs
 */
@Override
protected void doPost(HttpServletRequest request, HttpServletResponse response)
throws ServletException, IOException {
    processRequest(request, response);
```

```java
    }

    /**
     * Returns a short description of the servlet.
     * @return a String containing servlet description
     */
    @Override
    public String getServletInfo() {
        return "Short description";
    }// </editor-fold>

}


/*
 * To change this template, choose Tools | Templates
 * and open the template in the editor.
 */

package Servlet;

import Bean.expert;
import java.io.IOException;
import java.io.PrintWriter;
import java.sql.Connection;
import java.sql.ResultSet;
import java.sql.Statement;
import java.util.ArrayList;
import javax.servlet.RequestDispatcher;
import javax.servlet.ServletException;
import javax.servlet.annotation.WebServlet;
import javax.servlet.http.HttpServlet;
import javax.servlet.http.HttpServletRequest;
import javax.servlet.http.HttpServletResponse;
import util.DBUtil;

/**
 *
 * @author admin
 */
@WebServlet(name = "expert_select", urlPatterns = {"/expert_select"})
public class expert_select extends HttpServlet {

    /**
```

```java
 * Processes requests for both HTTP <code>GET</code> and <code>POST</code> methods.
 * @param request servlet request
 * @param response servlet response
 * @throws ServletException if a servlet-specific error occurs
 * @throws IOException if an I/O error occurs
 */
protected void processRequest(HttpServletRequest request, HttpServletResponse response)
throws ServletException, IOException {
    response.setContentType("text/html;charset=UTF-8");
    PrintWriter out = response.getWriter();
    try {
        /* TODO output your page here
        out.println("<html>");
        out.println("<head>");
        out.println("<title>Servlet expert_select</title>");
        out.println("</head>");
        out.println("<body>");
        out.println("<h1>Servlet expert_select at " + request.getContextPath() + "</h1>");
        out.println("</body>");
        out.println("</html>");
        */
    } finally {
        out.close();
    }
}

// <editor-fold defaultstate="collapsed" desc="HttpServlet methods. Click on the + sign on the left to edit the code.">
/**
 * Handles the HTTP <code>GET</code> method.
 * @param request servlet request
 * @param response servlet response
 * @throws ServletException if a servlet-specific error occurs
 * @throws IOException if an I/O error occurs
 */
@Override
protected void doGet(HttpServletRequest request, HttpServletResponse response)
throws ServletException, IOException {
    try {
        String proId = request.getParameter("proId");
        //获得数据库连接
```

```java
            DBUtil dbUtil = new DBUtil();
            //创建连接
            Connection con = dbUtil.getConnecetion();

            Statement stat = con.createStatement();
            //查询表
            ResultSet rs = stat.executeQuery("select * from dbo.T_expert where Expert_id in(select Expert_id from dbo.T_project_expert where Project_id = '" + proId.trim() + "')");

            ArrayList<expert> experts = new ArrayList<expert>();
            while(rs.next()){
            expert exp = new expert();
            exp.setExpert_id(rs.getString("Expert_id"));
            exp.setExpert_name(rs.getString("Expert_name"));
            exp.setDept(rs.getString("Dept"));
            exp.setSex(rs.getString("Sex"));
            exp.setID_Card(rs.getString("ID_Card"));
            experts.add(exp);
            }
            //关闭数据库的连接
            dbUtil.closeConnection();
             request.setAttribute("msg",experts);
            RequestDispatcher rd =
                    request.getRequestDispatcher("/jsp/expert_select.jsp");
                rd.forward(request, response);
                return;
        }catch(Exception ex){
        ex.printStackTrace();
        }
    }
}

/**
 * Handles the HTTP <code>POST</code> method.
 * @param request servlet request
 * @param response servlet response
 * @throws ServletException if a servlet-specific error occurs
 * @throws IOException if an I/O error occurs
 */
@Override
protected void doPost(HttpServletRequest request, HttpServletResponse response)
throws ServletException, IOException {

}
```

```
/**
 * Returns a short description of the servlet.
 * @return a String containing servlet description
 */
@Override
public String getServletInfo() {
    return "Short description";
}// </editor-fold>

}
```

任务四：建设工程监管信息系统交易流程模块

一、任务描述

你作为《建设工程监管信息系统》项目开发组的程序员，请实现如下功能：
- 交易流程步骤的列表显示；
- 交易流程步骤的添加和删除。

二、功能描述

（1）点击图4.1中的"交易流程步骤定义"菜单项，则在右边的主体部分中显示流程步骤列表。

图4.1 流程步骤定义页面

（2）对图4.1中的"流程步骤增加"的"＊"号部分进行必填并校验。
（3）点击图4.1中"确定"按钮，在流程步骤定义表中增加一条流程步骤信息。
（4）对图4.1中的"流程步骤删除"的"＊"号部分进行必填校验。
（5）点击图4.1中"删除"按钮，在流程步骤定义表中删除指定的流程步骤信息。
（6）在添加或删除流程步骤成功后，刷新页面中的"流程步骤列表"。
（7）测试程序，通过"流程步骤增加"页面增加两条以上流程步骤信息。

三、要求

1. 页面实现

以提供的素材为基础,实现图 4.1 所示页面。

2. 数据库实现

(1)创建数据库 ConstructionDB。

(2)创建流程步骤定义表(T_flow_step_def),表结构见表 4.1。

表 4.1 流程步骤定义表(T_flow_step_def)表结构

字段名	字段说明	字段类型	是否允许为空	备注
Step_no	流程步骤 ID	int	否	主键,从 1 开始自增
Step_name	流程步骤名称	varchar(32)	否	
Limit_time	时限	int	是	单位(天)
Step_des	流程步骤描述	varchar(64)	是	
URL	链接地址	varchar(64)	是	

(3)在表 T_flow_step_def 插入记录,见表 4.2。

表 4.2 流程步骤定义表(T_flow_step_def)记录

Step_no	Step_name	Limit_time	Step_des	URL
1	建设工程项目施工报建申请	10	施工报建	Flow/ConstructManager.html
2	入场交易登记	7	入场交易	Flow/ConstructManager.html

3. 功能实现

(1)功能需求如图 4.2 所示。

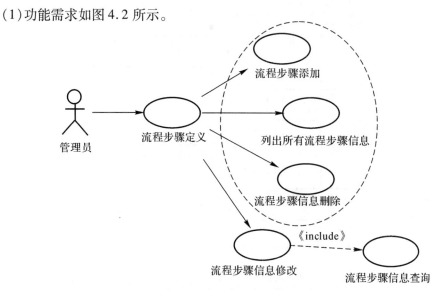

图 4.2 流程步骤定义模块用例图

（2）依据添加流程步骤信息活动图完成流程步骤添加功能，如图4.3所示。

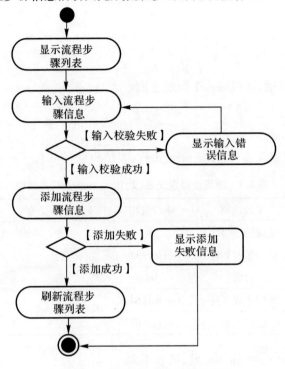

图 4.3　添加流程步骤信息活动图

（3）依据删除流程步骤信息活动图完成删除流程步骤功能，如图4.4所示。

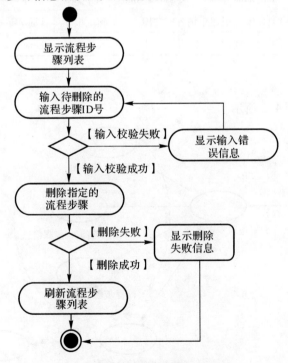

图 4.4　删除流程步骤信息活动图

四、必备知识

1. 数据库相关知识

（1）使用 MS SQL Server 2005/2008 创建数据库，创建数据表，设置表的字段，数据类型，主键，外键，约束。

（2）向数据表插入、删除、修改、查询数据。

2. 页面相关知识

（1）使用 HTML 制作项目页面。

（2）使用 CSS 控制页面的样式。

（3）使用 JavaScript 对页面必要的内容进行校验。

3. JSP 相关知识

（1）使用 JSTL 标准标签库控制页面显示逻辑。

（2）理解 JSP 的 request，response，session，application 的概念。

（3）使用 EL 表达式在页面显示数据。

（4）合理的使用转发和重定向控制项目的页面跳转。

（5）使用 JDBC 与数据库进行交互。

（6）MVC 模式下的分层架构，控制器，视图，模型的划分和通信。

五、解题思路

1. 数据库思路

（1）根据项目要求创建数据库和数据表，向数据表中插入合适的测试数据。

（2）导入 JDBC 驱动包，编写 JDBC 的连接工具代码。

（3）编写数据操作对象代码，负责进行与数据库的交互操作。

2. 视图层思路

（1）将提供的素材页面改写为 JSP 页面

（2）JSP 使用 JSTL 和 EL 负责控制页面显示的逻辑

3. 控制层思路

（1）使用 Servlet 类控制一次请求响应过程的处理。

（2）由 Servlet 按照顺序进行请求的处理、数据库交互、模型存取和封装、页面跳转逻辑控制等。

4. 模型层思路

使用 JavaBean 作为模型层，封装数据和行为。

六、操作步骤

1. 准备数据库

（1）根据项目要求，在 SQL Server2008 中创建 ConstructionDB 数据库、流程步骤定义表（T_flow_step_def），并插入测试数据。

（2）编写 JDBC 的连接工具代码。

```
package util;

import java.io.InputStream;
import java.sql.Connection;
import java.sql.DriverManager;
import java.util.Properties;

public class DBUtil {

    private Connection con;
    private String DRIVER;
    private String URL;
    private String DBNAME;
    private String DBPASS;

    public DBUtil() throws Exception {
        InputStream in =
            this.getClass().getClassLoader().getResourceAsStream("util/DBInfo.properties");
        if (null != in) {
            Properties props = new Properties();
            props.load(in);
            DRIVER = props.getProperty("driver");
            URL = props.getProperty("url");
            DBNAME = props.getProperty("user.name");
            DBPASS = props.getProperty("user.password");
        }
    }

    public Connection getConnecetion() throws Exception {
        Connection conn = null;
        try {
            if (this.con != null && !con.isClosed()) {
                conn = con;
            } else {
                Class.forName(DRIVER);
```

```java
            conn = DriverManager.getConnection(URL, DBNAME, DBPASS);
        }
    } catch (Exception ex) {
        System.out.println("数据库连接失败:" + ex);
        throw ex;
    }
    return conn;
}

public void closeConnection() throws Exception {
    try {
        if (this.con != null) {
            if (this.con.isClosed() == false) {
                this.con.close();
            }
        }
    } catch (Exception e) {
        System.out.println("关闭数据库失败:" + e);
        throw e;
    }
}
}
```

```java
/*
 * To change this template, choose Tools | Templates
 * and open the template in the editor.
 */
package dao;

import bean.StepBean;
import java.sql.SQLException;
import java.util.List;

/**
 *
 * @author Administrator
 */
public interface IStepOperator {

    public boolean insertStep(StepBean stepBean) throws SQLException;

    public List<StepBean> searchStep();
```

```java
    public StepBean searchStep(int step_no);

    public boolean updateStep(StepBean newStepBean) throws SQLException;

    public boolean deleteStep(int step_no) throws SQLException;
}

/*
 * To change this template, choose Tools | Templates
 * and open the template in the editor.
 */
package dao.impl;

import bean.StepBean;
import dao.IStepOperator;
import java.sql.Connection;
import java.sql.PreparedStatement;
import java.sql.ResultSet;
import java.sql.SQLException;
import java.util.ArrayList;
import java.util.logging.Level;
import java.util.logging.Logger;
import util.DBUtil;

/**
 *
 * @author Administrator
 */
public class StepOperatorImpl implements IStepOperator {

    @Override
    public boolean insertStep(StepBean stepBean) throws SQLException {
        boolean flag = false;
        try {
            DBUtil db = new DBUtil();
            Connection con = db.getConnecetion();
            String sql = "insert into T_flow_step_def values(?,?,?,?)";
            PreparedStatement ps = con.prepareStatement(sql);
            ps.setString(1, stepBean.getStep_name());
            ps.setInt(2, stepBean.getLimit_time());
            ps.setString(3, stepBean.getStep_des());
            ps.setString(4, stepBean.getURL());
            int result = ps.executeUpdate();
```

```java
                if (result > 0) {
                    flag = true;
                }
                db.closeConnection();
            } catch (Exception ex) {
                Logger.getLogger(StepOperatorImpl.class.getName()).log(Level.SEVERE, null, ex);
            }
            return flag;
        }


        @Override
        public ArrayList<StepBean> searchStep() {
            ArrayList<StepBean> list = new ArrayList<StepBean>();
            try {
                DBUtil db = new DBUtil();
                Connection con = db.getConnecetion();
                String sql = "select * from T_flow_step_def";
                PreparedStatement ps = con.prepareStatement(sql);
                ResultSet rs = ps.executeQuery();
                while (rs.next()) {
                    StepBean step = new StepBean();
                    step.setStep_no(rs.getInt(1));
                    step.setStep_name(rs.getString(2));
                    step.setLimit_time(rs.getInt(3));
                    step.setStep_des(rs.getString(4));
                    step.setURL(rs.getString(5));
                    list.add(step);
                }
                db.closeConnection();
            } catch (Exception ex) {
                Logger.getLogger(StepOperatorImpl.class.getName()).log(Level.SEVERE, null, ex);
            }
            return list;
        }


        @Override
        public boolean updateStep(StepBean newStepBean) throws SQLException {
            boolean flag = false;
            try {
                DBUtil db = new DBUtil();
                Connection con = db.getConnecetion();
                String sql = "update T_flow_step_def set Step_name = ?, Limit_time = ?, Step_des = ?, URL = ? where Step_no = ?";
```

```java
            PreparedStatement ps = con.prepareStatement(sql);
            ps.setString(1, newStepBean.getStep_name());
            ps.setInt(2, newStepBean.getLimit_time());
            ps.setString(3, newStepBean.getStep_des());
            ps.setString(4, newStepBean.getURL());
            ps.setInt(5, newStepBean.getStep_no());
            int result = ps.executeUpdate();
            if (result > 0) {
                flag = true;
            }
            db.closeConnection();
        } catch (Exception ex) {
Logger.getLogger(StepOperatorImpl.class.getName()).log(Level.SEVERE, null, ex);
        }
        return flag;
    }

    @Override
    public boolean deleteStep(int step_no) throws SQLException {
        boolean flag = false;
        try {
            DBUtil db = new DBUtil();
            Connection con = db.getConnecetion();
            String sql = "delete T_flow_step_def where Step_no = ?";
            PreparedStatement ps = con.prepareStatement(sql);
            ps.setInt(1, step_no);
            int result = ps.executeUpdate();
            if (result > 0) {
                flag = true;
            }
            db.closeConnection();
        } catch (Exception ex) {
Logger.getLogger(StepOperatorImpl.class.getName()).log(Level.SEVERE, null, ex);
        }
        return flag;
    }

    @Override
    public StepBean searchStep(int step_no) {
        StepBean step = null;
        try {
            DBUtil db = new DBUtil();
            Connection con = db.getConnecetion();
```

```java
            String sql = "select * T_flow_step_def where Step_no = ?";
            PreparedStatement ps = con.prepareStatement(sql);
            ps.setInt(1, step_no);
            ResultSet rs = ps.executeQuery();
            if (rs.next()) {
                step = new StepBean();
                step.setStep_no(rs.getInt(1));
                step.setStep_name(rs.getString(2));
                step.setLimit_time(rs.getInt(3));
                step.setStep_des(rs.getString(4));
                step.setURL(rs.getString(5));
            }
            db.closeConnection();
        } catch (Exception ex) {
            Logger.getLogger(StepOperatorImpl.class.getName()).log(Level.SEVERE, null, ex);
        }
        return step;
    }
}
```

2. 编写视图层代码

```jsp
<%--
    Document   : flow_def
    Created on : 2011-6-10, 20:20:38
    Author     : Administrator
--%>

<%@ page import="dao.impl.StepOperatorImpl"%>
<%@ page import="bean.StepBean"%>
<%@ page import="java.util.ArrayList"%>
<%@ page contentType="text/html" pageEncoding="UTF-8"%>
<html>
    <head>
        <meta http-equiv="Content-Type" content="text/html; charset=utf-8" />
        <title>无标题文档</title>
        <LINK href="css/content.css" type="text/css" rel="stylesheet" />
        <script type="text/javascript" src="js/step.js"></script>
    </head>
    <body>
        <table width="834" border="0" cellpadding="0" cellspacing="0">
            <tr>
                <td height="30" background="images/title01.jpg" class="title">&gt;&gt;
```

流程步骤定义</td>
　　　　　　　　　　　　　　<td width="27" height="30"><img src="images/title02.jpg" width="27" height="30" /></td>
　　　　　　　　　　　　　　<td height="30" bgcolor="#029AC5" class="txt"><span class="titletxt">您的位置：系统管理>流程步骤定义</span></td>
　　　　　　　　　　　　</tr>
　　　　　　　　　　</table>
　　　　　　　　　　
　　　　　　　　　　<table width="800" border="0" align="center">
　　　　　　　　　　　　<tr>
　　　　　　　　　　　　　　<td width="40%" rowspan="2" valign="top"><table width="100%" border="0" align="center" cellpadding="0" cellspacing="0">
　　　　　　　　　　　　　　　　<tr>
　　　　　　　　　　　　　　　　　　<td width="100%" height="30" bgcolor="#029AC5" class="titletxt">•流程步骤列表</td>
　　　　　　　　　　　　　　　　</tr>
　　　　　　　　　　　　　　　　<tr>
　　　　　　　　　　　　　　　　　　<td height="30" align="center"><form id="form2" name="form1" method="post" action="">
　　　　　　　　　　　　　　　　　　　　<table width="100%" border="1" align="center" cellpadding="0" cellspacing="0" bgcolor="#E7E7E7">
　　　　　　　　　　　　　　　　　　　　　　<tr>
　　　　　　　　　　　　　　　　　　　　　　　　<td width="23%" height="30" align="right" bgcolor="#80C6FF" class="txt">流程ID：</td>
　　　　　　　　　　　　　　　　　　　　　　　　<td width="77%" height="30" align="center" bgcolor="#80C6FF"><label for="textfield2" class="txt">流程名称</label></td>
　　　　　　　　　　　　　　　　　　　　　　</tr>
　　　　　　　　　　　　　　　　　　　　　　<%
　　　　　　　　　　　　　　　　　　　　　　　　ArrayList<StepBean> list = (ArrayList<StepBean>) request.getAttribute("list");
　　　　　　　　　　　　　　　　　　　　　　%>
　　　　　　　　　　　　　　　　　　　　　　<%
　　　　　　　　　　　　　　　　　　　　　　　　if (list == null) {
　　　　　　　　　　　　　　　　　　　　　　　　　　list = StepOperatorImpl.class.newInstance().searchStep();
　　　　　　　　　　　　　　　　　　　　　　　　}
　　　　　　　　　　　　　　　　　　　　　　　　for (int i = 0; list != null && i < list.size(); i++) {
　　　　　　　　　　　　　　　　　　　　　　　　　　StepBean step = (StepBean) list.get(i);
　　　　　　　　　　　　　　　　　　　　　　%>
　　　　　　　　　　　　　　　　　　　　　　<tr>
　　　　　　　　　　　　　　　　　　　　　　　　<td width="23%" height="30" align="

```
center" class="txt">
                              <%=step.getStep_no()%>
                        </td>
                        <td height="30" align="left" class="txt">
                              <%=step.getStep_name()%>
                        </td>
                        <%
                           }
                        %>
                      </table>
                    </form>
                  </td>
                </tr>
              </table></td>
              <td width="60%" valign="top"><table width="100%" border="0" align="center" cellpadding="0" cellspacing="0">
                <tr>
                  <td width="100%" height="30" bgcolor="#029AC5" class="titletxt">&#8226;流程步骤增加</td>
                </tr>
                <tr>
                  <td height="30" align="center" valign="top">
                    <form id="form1" name="form1" method="post" onsubmit="return vertify();" action="./InsertServlet">
                      <table width="100%" border="1" align="center" cellpadding="0" cellspacing="0" bgcolor="#E7E7E7">
                        <tr>
                          <td width="24%" height="30" align="right" class="txt">名称：</td>
                          <td height="30" align="left">
                            <input height="20" width="200" type="text" name="StepName" id="textfield2" onchange="checkName()" />
                            <span class="txtred">*</span>
                          </td>
                        </tr>
                        <tr>
                          <td height="30" align="right" class="txt">时限(天)：</td>
                          <td height="30" align="left">
                            <input height="20" width="50" type="text" name="limitTime" id="textfield5" onchange="checkLimitTime()" />
                            <span class="txtred">*</span>
```

```
            </td>
                    </tr>
                    <tr>
                        <td height="30" align="right" class="txt">描述:</td>
                        <td height="30" align="left">
                            <input height="20" width="200" type="text" name="des" id="textfield4" /></td>
                    </tr>
                    <tr>
                        <td height="30" align="right" class="txt">链接地址:</td>
                        <td height="30" align="left">
                            <input height="20" width="200" type="text" name="url" id="textfield4" /></td>
                    </tr>
                    <tr>
                        <td height="30" align="right" class="txt"> </td>
                        <td height="30" align="left">        
                            <input type="submit" name="button" id="button" value="确  定" /></td>
                    </tr>
                </table>
              </form>
            </td>
          </tr>
        </table></td>
      </tr>
      <tr>
        <td width="60%" height="1" valign="top"><table width="100%" border="0" align="center" cellpadding="0" cellspacing="0">
          <tr>
            <td width="100%" height="30" bgcolor="#029AC5" class="titletxt">&#8226;流程步骤删除</td>
          </tr>
          <tr>
            <td height="30" align="center"><form id="form3" name="form1" method="post" onsubmit="return checkNo();" action="./DeleteServlet">
              <table width="100%" border="1" align="center" cellpadding="0" cellspacing="0" bgcolor="#E7E7E7">
                <tr>
```

```html
                                            <td width="24%" height="30" align="right" class="txt">ID号:</td>
                                            <td height="30" align="left"><label for="textfield3"></label>
                                            <input height="20" width="50" type="text" name="stepNo" id="textfield3"/> <span class="txtred">*</span></td>
                                            <td align="left"><span class="txtred">
                                            <input type="submit" name="button3" id="button3" value="删  除"/>
                                            </span></td>
                                        </tr>
                                    </table>
                                </form></td>
                            </tr>
                        </table></td>
                </tr>
            </table>
            <p> </p>
        </body>
</html>
```

```jsp
<%--
    Document   : error
    Created on : 2011-6-11, 11:08:09
    Author     : Administrator
--%>

<%@ page contentType="text/html" pageEncoding="UTF-8"%>
<!DOCTYPE html>
<html>
    <head>
        <meta http-equiv="Content-Type" content="text/html; charset=UTF-8">
        <title>error info</title>
    </head>
    <body>
        <h1>
            <%=request.getAttribute("error")%>
        </h1>
    </body>
</html>
```

3. 编写模型层代码

```
/*
```

```java
 * To change this template, choose Tools | Templates
 * and open the template in the editor.
 */
package bean;

/**
 *
 * @author Administrator
 */
public class StepBean {

    private int step_no;
    private String step_name;
    private int limit_time;
    private String step_des;
    private String URL;

    /**
     * @return the step_no
     */
    public int getStep_no() {
        return step_no;
    }

    /**
     * @param step_no the step_no to set
     */
    public void setStep_no(int step_no) {
        this.step_no = step_no;
    }

    /**
     * @return the step_name
     */
    public String getStep_name() {
        return step_name;
    }

    /**
     * @param step_name the step_name to set
     */
    public void setStep_name(String step_name) {
        this.step_name = step_name;
```

```java
}

/**
 * @return the limit_time
 */
public int getLimit_time() {
    return limit_time;
}

/**
 * @param limit_time the limit_time to set
 */
public void setLimit_time(int limit_time) {
    this.limit_time = limit_time;
}

/**
 * @return the step_des
 */
public String getStep_des() {
    return step_des;
}

/**
 * @param step_des the step_des to set
 */
public void setStep_des(String step_des) {
    this.step_des = step_des;
}

/**
 * @return the URL
 */
public String getURL() {
    return URL;
}

/**
 * @param URL the URL to set
 */
public void setURL(String URL) {
    this.URL = URL;
}
```

}

4. 编写控制层代码

```
/*
 * To change this template, choose Tools | Templates
 * and open the template in the editor.
 */
package servlet;

import bean.StepBean;
import dao.IStepOperator;
import dao.impl.StepOperatorImpl;
import java.io.IOException;
import java.io.PrintWriter;
import java.sql.SQLException;
import java.util.logging.Level;
import java.util.logging.Logger;
import javax.servlet.ServletException;
import javax.servlet.http.HttpServlet;
import javax.servlet.http.HttpServletRequest;
import javax.servlet.http.HttpServletResponse;

/**
 *
 * @author Administrator
 */
public class InsertServlet extends HttpServlet {

    /**
     * Processes requests for both HTTP <code>GET</code> and <code>POST</code> methods.
     * @param request servlet request
     * @param response servlet response
     * @throws ServletException if a servlet-specific error occurs
     * @throws IOException if an I/O error occurs
     */
    protected void processRequest(HttpServletRequest request, HttpServletResponse response)
            throws ServletException, IOException {
        response.setContentType("text/html;charset=UTF-8");
        PrintWriter out = response.getWriter();
        try {
            String name = request.getParameter("StepName").trim();
            String limitTime = request.getParameter("limitTime").trim();
```

任务四:建设工程监管信息系统交易流程模块

```java
                String des = request.getParameter("des").trim();
                String URL = request.getParameter("url").trim();
                if ("".equals(name) || null == name) {
                    request.setAttribute("error", "名称不能为空!");
                    request.getRequestDispatcher("error.jsp").forward(request, response);
                }
                if ("".equals(limitTime) || null == limitTime) {
                    request.setAttribute("error", "时限不能为空!");
                    request.getRequestDispatcher("error.jsp").forward(request, response);
                }
                if (Integer.parseInt(limitTime) <= 0 || Integer.parseInt(limitTime) % 1 != 0) {
                    request.setAttribute("error", "时限必须大于零并且应为整数!");
                    request.getRequestDispatcher("error.jsp").forward(request, response);
                }
                StepBean step = new StepBean();
                step.setStep_name(name);
                step.setLimit_time(Integer.parseInt(limitTime));
                step.setStep_des(des);
                step.setURL(URL);
                IStepOperator so = StepOperatorImpl.class.newInstance();
                boolean result = so.insertStep(step);
                if (result) {
                    request.getRequestDispatcher("./SearchServlet").forward(request, response);
                } else {
                    request.setAttribute("error", "添加操作失败!");
                    request.getRequestDispatcher("error.jsp").forward(request, response);
                }
            } catch (SQLException ex) {
                Logger.getLogger(InsertServlet.class.getName()).log(Level.SEVERE, null, ex);
                request.setAttribute("error", "添加操作失败!");
                request.getRequestDispatcher("error.jsp").forward(request, response);
            } catch (InstantiationException ex) {
                Logger.getLogger(InsertServlet.class.getName()).log(Level.SEVERE, null, ex);
            } catch (IllegalAccessException ex) {
                Logger.getLogger(InsertServlet.class.getName()).log(Level.SEVERE, null, ex);
            } finally {
                out.close();
            }
        }

    // <editor-fold defaultstate="collapsed" desc="HttpServlet methods. Click on the + sign on the left to edit the code.">
    /**
```

```java
 * Handles the HTTP <code>GET</code> method.
 * @param request servlet request
 * @param response servlet response
 * @throws ServletException if a servlet-specific error occurs
 * @throws IOException if an I/O error occurs
 */
@Override
protected void doGet(HttpServletRequest request, HttpServletResponse response)
        throws ServletException, IOException {
    processRequest(request, response);
}

/**
 * Handles the HTTP <code>POST</code> method.
 * @param request servlet request
 * @param response servlet response
 * @throws ServletException if a servlet-specific error occurs
 * @throws IOException if an I/O error occurs
 */
@Override
protected void doPost(HttpServletRequest request, HttpServletResponse response)
        throws ServletException, IOException {
    processRequest(request, response);
}

/**
 * Returns a short description of the servlet.
 * @return a String containing servlet description
 */
@Override
public String getServletInfo() {
    return "Short description";
}// </editor-fold>
}

/*
 * To change this template, choose Tools | Templates
 * and open the template in the editor.
 */
package servlet;

import dao.IStepOperator;
import dao.impl.StepOperatorImpl;
```

任务四:建设工程监管信息系统交易流程模块

```java
import java.io.IOException;
import java.io.PrintWriter;
import java.sql.SQLException;
import java.util.logging.Level;
import java.util.logging.Logger;
import javax.servlet.ServletException;
import javax.servlet.http.HttpServlet;
import javax.servlet.http.HttpServletRequest;
import javax.servlet.http.HttpServletResponse;

/**
 *
 * @author Administrator
 */
public class DeleteServlet extends HttpServlet {

    /**
     * Processes requests for both HTTP <code>GET</code> and <code>POST</code> methods.
     * @param request servlet request
     * @param response servlet response
     * @throws ServletException if a servlet-specific error occurs
     * @throws IOException if an I/O error occurs
     */
    protected void processRequest(HttpServletRequest request, HttpServletResponse response)
            throws ServletException, IOException {
        response.setContentType("text/html;charset=UTF-8");
        PrintWriter out = response.getWriter();
        try {
            String stepNo = request.getParameter("stepNo").trim();
            if ("".equals(stepNo) || null == stepNo) {
                request.setAttribute("error", "步骤号不能为空!");
                request.getRequestDispatcher("error.jsp").forward(request, response);
            }
            if (Integer.parseInt(stepNo) <= 0 || Integer.parseInt(stepNo) % 1 != 0) {
                request.setAttribute("error", "步骤号必须大于零并且应为整数!");
                request.getRequestDispatcher("error.jsp").forward(request, response);
            }
            IStepOperator so = StepOperatorImpl.class.newInstance();
            boolean result = so.deleteStep(Integer.parseInt(stepNo));
            if (result) {
                request.getRequestDispatcher("./SearchServlet").forward(request, response);
            } else {
```

```java
                    request.setAttribute("error", "删除操作失败!");
request.getRequestDispatcher("error.jsp").forward(request, response);
                    }
            } catch (SQLException ex) {
Logger.getLogger(DeleteServlet.class.getName()).log(Level.SEVERE, null, ex);
            } catch (InstantiationException ex) {
Logger.getLogger(DeleteServlet.class.getName()).log(Level.SEVERE, null, ex);
            } catch (IllegalAccessException ex) {
Logger.getLogger(DeleteServlet.class.getName()).log(Level.SEVERE, null, ex);
            } finally {
                out.close();
            }
    }

    // <editor-fold defaultstate="collapsed" desc="HttpServlet methods. Click on the + sign on the left to edit the code.">
    /**
     * Handles the HTTP <code>GET</code> method.
     * @param request servlet request
     * @param response servlet response
     * @throws ServletException if a servlet-specific error occurs
     * @throws IOException if an I/O error occurs
     */
    @Override
    protected void doGet(HttpServletRequest request, HttpServletResponse response)
            throws ServletException, IOException {
        processRequest(request, response);
    }

    /**
     * Handles the HTTP <code>POST</code> method.
     * @param request servlet request
     * @param response servlet response
     * @throws ServletException if a servlet-specific error occurs
     * @throws IOException if an I/O error occurs
     */
    @Override
    protected void doPost(HttpServletRequest request, HttpServletResponse response)
            throws ServletException, IOException {
        processRequest(request, response);
    }

    /**
```

* Returns a short description of the servlet.
 * @return a String containing servlet description
 */
@Override
public String getServletInfo() {
    return "Short description";
}// </editor-fold>
}

/*
 * To change this template, choose Tools | Templates
 * and open the template in the editor.
 */
package servlet;

import bean.StepBean;
import dao.IStepOperator;
import dao.impl.StepOperatorImpl;
import java.io.IOException;
import java.io.PrintWriter;
import java.util.List;
import java.util.logging.Level;
import java.util.logging.Logger;
import javax.servlet.ServletException;
import javax.servlet.http.HttpServlet;
import javax.servlet.http.HttpServletRequest;
import javax.servlet.http.HttpServletResponse;

/**
 *
 * @author Administrator
 */
public class SearchServlet extends HttpServlet {

    /**
     * Processes requests for both HTTP <code>GET</code> and <code>POST</code> methods.
     * @param request servlet request
     * @param response servlet response
     * @throws ServletException if a servlet-specific error occurs
     * @throws IOException if an I/O error occurs
     */
    protected void processRequest(HttpServletRequest request, HttpServletResponse response)

```java
        throws ServletException, IOException, InstantiationException, IllegalAccessException {
    response.setContentType("text/html;charset=UTF-8");
    PrintWriter out = response.getWriter();
    try {
        IStepOperator so = StepOperatorImpl.class.newInstance();
        List<StepBean> list = so.searchStep();
        request.setAttribute("list", list);
        request.getRequestDispatcher("flow_def.jsp").forward(request, response);
    } finally {
        out.close();
    }
}


// <editor-fold defaultstate="collapsed" desc="HttpServlet methods. Click on the + sign on the left to edit the code.">
/**
 * Handles the HTTP <code>GET</code> method.
 * @param request servlet request
 * @param response servlet response
 * @throws ServletException if a servlet-specific error occurs
 * @throws IOException if an I/O error occurs
 */
@Override
protected void doGet(HttpServletRequest request, HttpServletResponse response)
        throws ServletException, IOException {
    try {
        processRequest(request, response);
    } catch (InstantiationException ex) {
        Logger.getLogger(SearchServlet.class.getName()).log(Level.SEVERE, null, ex);
    } catch (IllegalAccessException ex) {
        Logger.getLogger(SearchServlet.class.getName()).log(Level.SEVERE, null, ex);
    }
}


/**
 * Handles the HTTP <code>POST</code> method.
 * @param request servlet request
 * @param response servlet response
 * @throws ServletException if a servlet-specific error occurs
 * @throws IOException if an I/O error occurs
 */
@Override
protected void doPost(HttpServletRequest request, HttpServletResponse response)
```

```java
            throws ServletException, IOException {
        try {
            processRequest(request, response);
        } catch (InstantiationException ex) {
            Logger.getLogger(SearchServlet.class.getName()).log(Level.SEVERE, null, ex);
        } catch (IllegalAccessException ex) {
            Logger.getLogger(SearchServlet.class.getName()).log(Level.SEVERE, null, ex);
        }
    }

    /**
     * Returns a short description of the servlet.
     * @return a String containing servlet description
     */
    @Override
    public String getServletInfo() {
        return "Short description";
    }// </editor-fold>
}

/*
 * To change this template, choose Tools | Templates
 * and open the template in the editor.
 */
package servlet;

import java.io.IOException;
import java.io.PrintWriter;
import javax.servlet.ServletException;
import javax.servlet.http.HttpServlet;
import javax.servlet.http.HttpServletRequest;
import javax.servlet.http.HttpServletResponse;

/**
 *
 * @author Administrator
 */
public class UpdateServlet extends HttpServlet {

    /**
     * Processes requests for both HTTP <code>GET</code> and <code>POST</code> methods.
     * @param request servlet request
```

```java
 * @param response servlet response
 * @throws ServletException if a servlet-specific error occurs
 * @throws IOException if an I/O error occurs
 */
protected void processRequest(HttpServletRequest request, HttpServletResponse response)
        throws ServletException, IOException {
    response.setContentType("text/html;charset=UTF-8");
    PrintWriter out = response.getWriter();
    try {
        /* TODO output your page here
        out.println("<html>");
        out.println("<head>");
        out.println("<title>Servlet UpdateServlet</title>");
        out.println("</head>");
        out.println("<body>");
        out.println("<h1>Servlet UpdateServlet at " + request.getContextPath() + "</h1>");
        out.println("</body>");
        out.println("</html>");
        */
    } finally {
        out.close();
    }
}

// <editor-fold defaultstate="collapsed" desc="HttpServlet methods. Click on the + sign on the left to edit the code.">
/**
 * Handles the HTTP <code>GET</code> method.
 * @param request servlet request
 * @param response servlet response
 * @throws ServletException if a servlet-specific error occurs
 * @throws IOException if an I/O error occurs
 */
@Override
protected void doGet(HttpServletRequest request, HttpServletResponse response)
        throws ServletException, IOException {
    processRequest(request, response);
}

/**
 * Handles the HTTP <code>POST</code> method.
 * @param request servlet request
```

```
 * @param response servlet response
 * @throws ServletException if a servlet-specific error occurs
 * @throws IOException if an I/O error occurs
 */
@Override
protected void doPost(HttpServletRequest request, HttpServletResponse response)
        throws ServletException, IOException {
    processRequest(request, response);
}

/**
 * Returns a short description of the servlet.
 * @return a String containing servlet description
 */
@Override
public String getServletInfo() {
    return "Short description";
}// </editor-fold>

}
```

任务五：电子商务购物网站产品查询模块

一、任务描述

你作为《电子商务购物网站》项目开发组的程序员，请实现如下功能：
➢ 按产品名称查询产品数据。

二、功能描述

(1)进入图 5.1 所示页面时,默认显示全部产品信息。

(2)在图 5.1 中产品查询功能为模糊查询,例如:输入"富士相机"或者"相机"均能查询结果,如图 5.2 所示;当查询失败时,则显示"对不起,没有你查询的产品,请更换条件重新查询"提示,如图 5.3 所示。

(3)产品名称为空时,点击"查询产品"按钮将显示全部产品信息。

图 5.1 产品信息查询页面

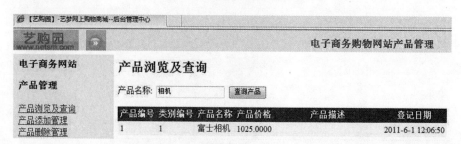

图 5.2 产品信息查询成功后的显示页面

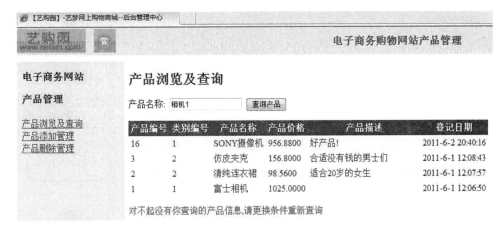

图 5.3　产品信息查询失败后的显示页面

三、要求

1. 页面实现

以提供的素材为基础,实现图 5.1 所示页面。

2. 数据库实现

(1)创建数据库 ProductDB。

(2)创建产品类别表(T_category),表结构见表 5.1。

表 5.1　产品类别表(T_category)表结构

字段名	字段说明	字段类型	是否允许为空	备注
Category_id	产品类别 ID	int	否	Pk(主键)
Category_name	产品类别名称	varchar(30)	否	
Register_date	默认值为当前录入时间	datetime	否	日期型

(3)创建产品表(T_product),表结构见表 5.2。

表 5.2　产品表(T_product)表结构

字段名	字段说明	字段类型	是否允许为空	备注
Product_id	产品编号	int	否	Pk(主键) 标识列
Category_id	产品类别 ID	int	否	FK(外键)
Product_name	产品名称	varchar(50)	否	
Price	产品价格	money	否	货币型
Remark	产品描述	varchar(2000)	否	
Register_date	默认值为当前录入时间	datetime	否	日期型

3. 功能实现

(1)新建一个 web 项目,该项目名称为 ProductAdmin。

(2)实现按产品名称查询产品数据的功能。要求在产品显示页面中,产品信息列表的表头均为中文。

(3)无关键字的查询,显示全部的产品信息,操作流程如图 5.4 所示。

(4)实现模糊查询。

(5)查询失败时,给出提示。

四、必备知识

图 5.4　按产品名称查询产品活动图

1. 数据库相关知识

(1)使用 MS SQL Server 2005/2008 创建数据库,创建数据表,设置表的字段,数据类型,主键,外键,约束。

(2)向数据表插入、删除、修改、查询数据。

2. 页面相关知识

(1)使用 HTML 制作项目页面。

(2)使用 CSS 控制页面的样式。

(3)使用 JavaScript 对页面必要的内容进行校验。

3. JSP 相关知识

(1)使用 JSTL 标准标签库控制页面显示逻辑。

(2)理解 JSP 的 request,response,session,application 的概念。

(3)使用 EL 表达式在页面显示数据。

(4)合理地使用转发和重定向控制项目的页面跳转。

(5)使用 JDBC 与数据库进行交互。

(6)MVC 模式下的分层架构,控制器,视图,模型的划分和通信。

五、解题思路

1. 数据库思路

(1)根据项目要求创建数据库和数据表,向数据表中插入合适的测试数据。

(2)导入 JDBC 驱动包,编写 JDBC 的连接工具代码。

(3)编写数据操作对象代码,负责进行与数据库的交互操作。

2. 视图层思路

(1)将提供的素材页面改写为 JSP 页面。

(2)JSP 使用 JSTL 和 EL 负责控制页面显示的逻辑。

3. 控制层思路

（1）使用 Servlet 类控制一次请求响应过程的处理。

（2）由 Servlet 按照顺序进行请求的处理、数据库交互、模型存取和封装、页面跳转逻辑控制等。

4. 模型层思路

使用 JavaBean 作为模型层，封装数据和行为。

六、操作步骤

1. 准备数据库

（1）根据项目要求，在 SQL Server2008 中创建 ProductDB 数据库、产品类别表（T_category）、产品表（T_product），并插入测试数据。

（2）编写 JDBC 的连接工具代码。

```java
package com.antonio.util;

import java.sql.Connection;
import java.sql.DriverManager;
import java.sql.SQLException;

public class JDBCUtils {
    private static final String DRIVER = "com.microsoft.sqlserver.jdbc.SQLServerDriver";
    private static final String URL = "jdbc:sqlserver://127.0.0.1:1433;DatabaseName=ProductDB";
    private static final String USER = "sa";
    private static final String PASS = "1234";
    private static Connection connection = null;
    public static Connection getConnection() {

        try {
            Class.forName(DRIVER);
            connection = DriverManager.getConnection(URL, USER, PASS);
            System.out.println("数据库连接成功");

        } catch (ClassNotFoundException e) {
            System.out.println("找不到数据库驱动");
        } catch (SQLException e) {
            System.out.println("数据库连接失败");
        }

        return connection;
    }
```

}

2. 编写视图层代码

```jsp
<%@ page pageEncoding="utf-8" %>
<%@ page import="java.util.List" %>
<%@ page import="com.antonio.bean.Product" %>
<!DOCTYPE html PUBLIC "-//W3C//DTD XHTML 1.0 Transitional//EN" "http://www.w3.org/TR/xhtml1/DTD/xhtml1-transitional.dtd">
<html xmlns="http://www.w3.org/1999/xhtml">
<head>
    <title>产品浏览及查询</title>
</head>
<body>
    <h2>产品浏览及查询</h2>
    <div>
        产品名称: 
        <form action="findProduct">
            <input name="txtPname" type="text" id="txtPname" style="width:127px;" /> 
            <input type="submit" name="Button1" value="查询产品" id="Button1" /><br />
        </form>
    </div>
    <p></p>
    <div>
        <div>
            <table cellspacing="0" cellpadding="4" border="0" style="color:#333333;border-collapse:collapse;">
                <tr style="color:White;background-color:#990000;font-weight:bold;">
                    <th>产品编号</th>
                    <th>类别编号</th>
                    <th>产品名称</th>
                    <th>产品价格</th>
                    <th>产品描述</th>
                    <th>登记日期</th>
                </tr>
                <%
                    List<Product> productList = (List<Product>)session.getAttribute("productList");
                    if(productList.size()<=0){
                %>
                <tr style="color:#333333;background-color:#FFFBD6;">
```

```jsp
            <td colspan="6">对不起,当前没有可查询的产品</td>
        </tr>
<%
    }else{
        for(Product product : productList){
%>
        <tr style="color:#333333;background-color:#FFFBD6;">
            <td><%=product.getProduct_id() %></td>
            <td><%=product.getCategory_id() %></td>
            <td><%=product.getProduct_name() %></td>
            <td><%=product.getPrice() %></td>
            <td style="width:200px;"><%=product.getRemark() %></td>
            <td><%=product.getRegister_date() %></td>
        </tr>
<% }
    } %>
    </table>
</div>

</div>

</body>
</html>
```

3. 编写模型层代码

```java
package com.antonio.bean;

import java.sql.Timestamp;

public class Product {
    private Integer Product_id;
    private Integer Category_id;
    private String Product_name;
    private double Price;
    private String Remark;
    private Timestamp Register_date;
    public Product(){
        super();
    }
    public Product(Integer product_id, Integer category_id,
            String product_name, double price, String remark,
            Timestamp register_date){
```

```java
        super();
        Product_id = product_id;
    Category_id = category_id;
    Product_name = product_name;
    Price = price;
    Remark = remark;
    Register_date = register_date;
}
public Integer getProduct_id() {
    return Product_id;
}
public void setProduct_id(Integer product_id) {
    Product_id = product_id;
}
public Integer getCategory_id() {
    return Category_id;
}
public void setCategory_id(Integer category_id) {
    Category_id = category_id;
}
public String getProduct_name() {
    return Product_name;
}
public void setProduct_name(String product_name) {
    Product_name = product_name;
}
public double getPrice() {
    return Price;
}
public void setPrice(double price) {
    Price = price;
}
public String getRemark() {
    return Remark;
}
public void setRemark(String remark) {
    Remark = remark;
}
public Timestamp getRegister_date() {
    return Register_date;
}
public void setRegister_date(Timestamp register_date) {
    Register_date = register_date;
```

                }

        }

4. 编写控制层代码

```java
package com.antonio.servlet;

import java.io.IOException;
import java.sql.Connection;
import java.sql.ResultSet;
import java.sql.SQLException;
import java.sql.Statement;
import java.util.ArrayList;
import java.util.List;

import javax.servlet.ServletException;
import javax.servlet.http.HttpServlet;
import javax.servlet.http.HttpServletRequest;
import javax.servlet.http.HttpServletResponse;

import com.antonio.bean.Product;
import com.antonio.util.JDBCUtils;

public class GetAllServlet extends HttpServlet{

    @Override
    protected void service(HttpServletRequest request, HttpServletResponse response)
            throws ServletException, IOException {
        //连接数据库
        Connection connection = JDBCUtils.getConnection();
        //查询所有商品
        try {
            Statement stmt = connection.createStatement();
            String sql = "select * from T_product;";
            ResultSet rs = stmt.executeQuery(sql);
            List<Product> productList = new ArrayList<Product>();
            while(rs.next()){
                Product product = new Product();
                product.setProduct_id(rs.getInt("Product_id"));
                product.setCategory_id(rs.getInt("Category_id"));
                product.setProduct_name(rs.getString("Product_name"));
                product.setPrice(rs.getDouble("Price"));
                product.setRemark(rs.getString("Remark"));
```

```java
                product.setRegister_date(rs.getTimestamp("Register_date"));
                productList.add(product);
            }

            //将商品列表放入 session
            request.getSession().setAttribute("productList", productList);

            //转发到 index.jsp
            request.getRequestDispatcher("/index.jsp").forward(request, response);

        } catch (SQLException e) {
            // TODO 自动生成的 catch 块
            e.printStackTrace();
        }
    }
}


package com.antonio.servlet;

import java.io.IOException;
import java.sql.Connection;
import java.sql.PreparedStatement;
import java.sql.ResultSet;
import java.sql.SQLException;
import java.util.ArrayList;
import java.util.List;

import javax.servlet.ServletException;
import javax.servlet.http.HttpServlet;
import javax.servlet.http.HttpServletRequest;
import javax.servlet.http.HttpServletResponse;

import com.antonio.bean.Product;
import com.antonio.util.JDBCUtils;

public class FindProductServlet extends HttpServlet{

    @Override
    protected void service(HttpServletRequest request, HttpServletResponse response)
            throws ServletException, IOException {
        //从请求中获取关键字参数
        String productName = new String(request.getParameter("txtPname").getBytes("iso-8859-
```

```java
1"),"utf-8");
        //从数据库进行模糊查询
        //连接数据库
        Connection connection = JDBCUtils.getConnection();
        //查询所有商品
        try {
            String sql = "select * from T_product where Product_name like ?;";
            PreparedStatement stmt = connection.prepareStatement(sql);
            stmt.setString(1, "%" + productName + "%");
            ResultSet rs = stmt.executeQuery();

            List<Product> productList = new ArrayList<Product>();
            while(rs.next()){
                Product product = new Product();
                product.setProduct_id(rs.getInt("Product_id"));
                product.setCategory_id(rs.getInt("Category_id"));
                product.setProduct_name(rs.getString("Product_name"));
                product.setPrice(rs.getDouble("Price"));
                product.setRemark(rs.getString("Remark"));
                product.setRegister_date(rs.getTimestamp("Register_date"));
                productList.add(product);
            }
            if(productList.size()<=0){
                /*去查询全部产品*/
                /*转发到getAll*/
                request.getRequestDispatcher("/getAll").forward(request, response);
            }else{

                //将商品列表放入session
                request.getSession().setAttribute("productList", productList);

                //转发到index.jsp
        request.getRequestDispatcher("/ProductList.jsp").forward(request, response);


            }
        } catch (SQLException e) {
            // TODO 自动生成的 catch 块
            e.printStackTrace();
        }
    }
}
```

任务六：建设用地审批电子报盘管理系统行政区划模块

一、任务描述

你作为《建设用地审批电子报盘管理系统》项目开发组的程序员，请实现如下功能：
➢ 行政区划信息的列表显示；
➢ 行政区划信息的添加。

二、功能描述

（1）点击图6.1中左边导航条中的"行政区划"菜单项，则在右边的主体部分显示行政区划信息列表。

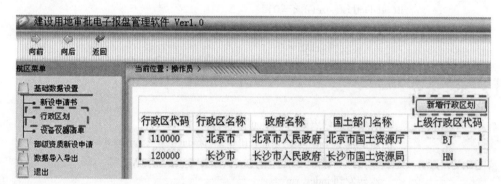

图6.1　行政区划信息列表页面

（2）点击图6.1中的"新增行政区划"按钮，则进入行政区划信息录入页面，如图6.2所示。
（3）对图6.2中打"＊"号的输入部分进行必填校验。
（4）点击"确定"按钮，在行政区划信息表中增加一条行政区划信息。
（5）行政区划信息增加成功后，自动定位到行政区划信息列表页面，显示更新后的行政区划信息列表，如图6.1。
（6）测试程序，通过行政区划信息录入页面增加两条以上行政区划信息。

任务六：建设用地审批电子报盘管理系统行政区划模块

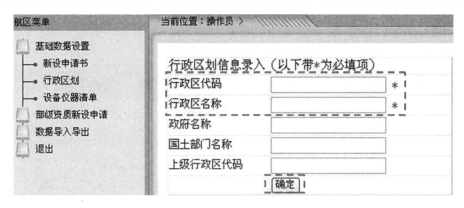

图 6.2　行政区划信息录入页面

三、要求

1. 界面实现

以提供的素材为基础，实现图 6.1、图 6.2 所示页面。

2. 数据库实现

（1）创建数据库 LandDB。

（2）创建行政区划信息表（T_bl_canton_code），表结构见表 6.1。

表 6.1　行政区划信息表（T_bl_canton_code）表结构

字段名	字段说明	字段类型	允许为空	备注
Cin_code	行政区代码	Int	否	主键
Cin_name	行政区名称	varchar(60)	否	
Gov_name	政府名称	varchar(60)	是	
Land_dp_name	国土部门名称	varchar(60)	是	
Parent_code	上级行政区代码	varchar(20)	是	

（3）在表 T_bl_canton_code 中插入记录，见表 6.2。

表 6.2　行政区划信息表（T_bl_canton_code）记录

Cin_code	Cin_name	Gov_name	Land_dp_name	Parent_code
110000	北京市	北京市人民政府	北京市国土资源厅	BJ
120000	长沙市	长沙市人民政府	长沙市国土资源局	HN

3. 功能实现

（1）功能需求如图 6.3 所示。

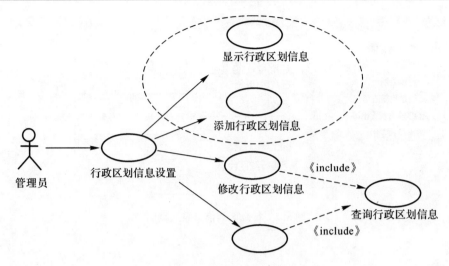

图6.3 行政区划信息设置模块用例图

（2）依据行政区划信息列表活动图完成行政区划信息列表显示功能，如图6.4所示。

（3）依据添加行政区划信息活动图完成添加行政区划信息功能，如图6.5所示。

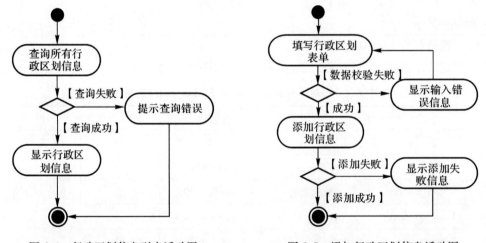

图6.4 行政区划信息列表活动图　　　　图6.5 添加行政区划信息活动图

四、必备知识

1. 数据库相关知识

（1）使用 MS SQL Server 2005/2008 创建数据库，创建数据表，设置表的字段，数据类型，主键，外键，约束。

（2）向数据表插入、删除、修改、查询数据。

2. 页面相关知识

（1）使用 HTML 制作项目页面。

（2）使用 CSS 控制页面的样式。

(3)使用 JavaScript 对页面必要的内容进行校验。

3. JSP 相关知识

(1)使用 JSTL 标准标签库控制页面显示逻辑。
(2)理解 JSP 的 request、response、session、application 的概念。
(3)使用 EL 表达式在页面显示数据。
(4)合理地使用转发和重定向控制项目的页面跳转。
(5)使用 JDBC 与数据库进行交互。
(6)MVC 模式下的分层架构,控制器,视图,模型的划分和通信。

五、解题思路

1. 数据库思路

(1)根据项目要求创建数据库和数据表,向数据表中插入合适的测试数据。
(2)导入 JDBC 驱动包,编写 JDBC 的连接工具代码。
(3)编写数据操作对象代码,负责进行与数据库的交互操作。

2. 视图层思路

(1)将提供的素材页面改写为 JSP 页面。
(2)JSP 使用 JSTL 和 EL 负责控制页面显示的逻辑。

3. 控制层思路

(1)使用 Servlet 类控制一次请求响应过程的处理。
(2)由 Servlet 按照顺序进行请求的处理、数据库交互、模型存取和封装、页面跳转逻辑控制等。

4. 模型层思路

使用 JavaBean 作为模型层,封装数据和行为。

六、操作步骤

1. 准备数据库

(1)根据项目要求,在 SQL Server2008 中创建 LandDB 数据库、行政区划信息表(T_bl_canton_code),并插入测试数据。
(2)编写 JDBC 的连接工具代码:

```
package com.dao;

import java.sql.Connection;
import java.sql.DriverManager;
import java.sql.PreparedStatement;
import java.sql.ResultSet;
```

```java
import java.sql.SQLException;

/**
 * 数据库连接类
 * @author 唐伟
 *
 */
public class DBConnection {
    private static final String DRIVER = "com.microsoft.sqlserver.jdbc.SQLServerDriver";
    private static final String URL = "jdbc:sqlserver://localhost:1433;DatabaseName=LandDB";
    private static final String USER = "sa";
    private static final String PWD = "123";

    private static Connection con;
    private static PreparedStatement prst;
    private static ResultSet rest;

    /**
     * 加载驱动
     */
    static {
        try {
            Class.forName(DRIVER);
        } catch (ClassNotFoundException e) {
            e.printStackTrace();
        }
    }

    /**
     * 取得连接
     * @return
     */
    public static Connection getConnection() {
        try {
            con = DriverManager.getConnection(URL, USER, PWD);
        } catch (SQLException e) {
            e.printStackTrace();
        }
        return con;
    }

    /**
     * 关闭连接
```

```java
         */
        public static void close() {
            try {
                if(rest! = null) {
                    rest.close();
                }
                if(prst! = null) {
                    prst.close();
                }
                if(con! = null) {
                    con.close();
                }
            } catch (SQLException e) {
                e.printStackTrace();
            }
        }
    }


package com.dao;

import java.util.List;

import com.entity.Canton;

/**
 * 行政区划操作接口
 * @author 唐伟
 *
 */
public interface CantonDao {
    /**
     * 添加行政区划信息
     * @param canton
     * @return
     */
    boolean addCanton(Canton canton);

    /**
     * 修改行政区划信息
     * @param canton
     * @return
     */
    boolean updateCanton(Canton canton);
```

```java
/**
 * 删除行政区划信息
 * @param id
 * @return
 */
boolean deleteCanton(int id);

/**
 * 查询所有的行政区划信息
 * @return
 */
List<Canton> listCanton();

/**
 * 查询行政区划信息
 * @param id
 * @return
 */
Canton findCanton(int id);
}


package com.dao;

import java.sql.Connection;
import java.sql.PreparedStatement;
import java.sql.ResultSet;
import java.sql.SQLException;
import java.util.ArrayList;
import java.util.List;

import com.entity.Canton;

/**
 * 行政区划操作实现类
 * @author 唐伟
 *
 */
public class CantonDaoImpl implements CantonDao {

    public boolean addCanton(Canton canton) {
        boolean flag = false;
        String sql = "insert into T_BL_CANTON_CODE values(?,?,?,?,?)";
```

```java
        Connection con = null;
        PreparedStatement prst = null;
        con = DBConnection.getConnection();
        try {
            prst = con.prepareStatement(sql);
            prst.setInt(1, canton.getCtn_code());
            prst.setString(2, canton.getCtn_name());
            prst.setString(3, canton.getGov_name());
            prst.setString(4, canton.getLand_dp_name());
            prst.setInt(5, canton.getParent_code());
            if(prst.executeUpdate() != -1) {
                flag = true;
            }
        } catch (SQLException e) {
            e.printStackTrace();
        } finally {
            DBConnection.close();
        }
        return flag;
    }

    public boolean deleteCanton(int id) {
        // TODO Auto-generated method stub
        return false;
    }

    public Canton findCanton(int id) {
        // TODO Auto-generated method stub
        return null;
    }

    public List<Canton> listCanton() {
        List<Canton> list = new ArrayList<Canton>();
        String sql = "select * from T_BL_CANTON_CODE";
        Connection con = null;
        PreparedStatement prst = null;
        ResultSet rest = null;
        con = DBConnection.getConnection();
        try {
            prst = con.prepareStatement(sql);
            rest = prst.executeQuery();
            while(rest.next()) {
                Canton canton = new Canton();
```

```java
            canton.setCtn_code(rest.getInt("ctn_code"));
            canton.setCtn_name(rest.getString("ctn_name"));
            canton.setGov_name(rest.getString("gov_name"));
            canton.setLand_dp_name(rest.getString("land_dp_name"));
            canton.setParent_code(rest.getInt("parent_code"));
            list.add(canton);
        }
    } catch (SQLException e) {
        e.printStackTrace();
    }
    return list;
}

public boolean updateCanton(Canton canton) {
    // TODO Auto-generated method stub
    return false;
}

}
```

2. 编写视图层代码

```jsp
<%@ page language="java" import="java.util.*" pageEncoding="gb2312"%>
<%
String path = request.getContextPath();
String basePath = request.getScheme()+"://"+request.getServerName()+":"+request.getServerPort()+path+"/";
%>

<!DOCTYPE HTML PUBLIC "-//W3C//DTD HTML 4.01 Transitional//EN">
<html>
  <head>
    <base href="<%=basePath%>">

    <title>My JSP 'manager.jsp' starting page</title>

    <meta http-equiv="pragma" content="no-cache">
    <meta http-equiv="cache-control" content="no-cache">
    <meta http-equiv="expires" content="0">
    <meta http-equiv="keywords" content="keyword1,keyword2,keyword3">
    <meta http-equiv="description" content="This is my page">
    <link href="css/style.css" rel=stylesheet type=text/css>
    <script language="VBScript">
    <!--
```

```
Sub GoForward()
    history.go(1)
End Sub

Sub GoBack()
    history.go(-1)
End Sub
-->
</script>

</head>

<body topmargin='0' leftmargin='0' background='images/bodybackground.jpg'>
<table border="0" cellpadding="0" cellspacing="0" background='images/header.jpg' align='center' valign='top' width="100%" height="77">
<tr>
<td style="background-image:url('images/headerleft.jpg');background-repeat:no-repeat" width='35' height='24'>
     
</td>
<td align='left' width='100%'>
<font style='font-size:12pt;color:#000000;'>建设用地审批电子报盘管理软件 Ver1.0</font>
</td>
</tr>
<tr>
<td height='50' colspan='2' width='100%'>
<table border="0" cellspacing="0">
<tr>
<td width="16"> </td>
<td width="50"><div align="center"><img src="images/back.gif" width="20" height="20" onClick="GoBack"></div></td>
<td width="50"><div align="center"><img src="images/next.gif" width="20" height="20" onClick="GoForward"></div></td>
<td width="50"><div align="center"><a href="../console/index.html" target="_top"><img src="images/return.gif" width="20" height="20" border="0"></a></div></td>
<td width="16"> </td>
<td width="68"> </td>
<td width="68"> </td>
<td width="68"> </td>
<td width="68"> </td>
<td width="68"> </td>
<td width="68"> </td>
```

```
            < td width = "68" >   </td >
            < td width = "68" >   </td >
            < td width = "68" >   </td >
            < td width = "140" >   </td >
        </tr >
        < tr >
            < td >   </td >
            < td > < div align = "center" > < font color = "#3074A2" style = "font - size:9pt;color:#000000;" > 向前 </font > </div > </td >
            < td > < div align = "center" > < font color = "#3074A2" style = "font - size:9pt;color:#000000" > 向后 </font > </div > </td >
            < td > < div align = "center" > < font color = "#3074A2" style = "font - size:9pt;color:#000000" > 返回 </font > </div > </td >
        </tr >
    </table >
  </td >
</tr >
< tr >
  < td background = 'images/bodybackground.jpg' width = '100%' nowrap colspan = '2' >
  < table width = '100%' cellpadding = "0" cellspacing = "0" border = '0' >
  < tr >
    < td background = 'images/bodybackground.jpg' nowrap width = '186' >
        < font style = 'font - size:9pt' > 导航区菜单 </font >

    </td >
    < td width = '100%' align = 'right' height = '23' >
        < iFRAME width = '100%' height = '100%'  scrolling = 'no' frameborder = '0' NAME = "CurSite" SRC = "current_site.jsp" > </iframe >
    </td >
  </tr >
  </table >
  </td >
</tr >
</table >
< table width = "100%" cellpadding = "0" cellspacing = "0" height = "500" align = "center" topmargin = '0' border = '0' >
    < tr >
        < td width = "1" background = 'images/bodybackground.jpg' cellpadding = "0" cellspacing = "0" >  

        </td >
        < td valign = 'top' align = 'left' name = "frmTitle1" id = frmTitle1 background = "images/bodybackground.jpg" style = "width:177" >
```

```
            <iFRAME width='100%' height='100%' scrolling='auto' frameborder='0' NAME="index_tree"
SRC="index_tree1.jsp" style="HEIGHT:100%;VISIBILITY:inherit;WIDTH:177px;Z-INDEX:
2"></iframe>
          </td>
          <TD background='images/bodybackground.jpg' width="0">
            <TABLE border=0 cellPadding=0 cellSpacing=0>
              <TBODY>
                <TR>
                  <TD onclick=switchSysBar() style="HEIGHT:100%;WIDTH:100%">
          <SPAN class=navPoint id=switchPoint title=关闭/打开左栏></SPAN>
                  </TD>
                </TR>
              </TBODY>
            </TABLE>
          </TD>
          <TD width='1'>
            <TABLE border=0 cellPadding=0 cellSpacing=0 height='100%' width="16">
              <TR>
                <TD background='images/main_left_top.jpg' height='16' width='16'>
                </TD>
              </TR>
              <TR>
                <TD background='images/main_left_middle.jpg' height='100%' width='16'>
                </TD>
              </TR>
              <TR>
                <TD background='images/main_left_bottom.jpg' height='24' width='16'>
                </TD>
              </TR>
            </TABLE>
          </TD>
          <TD width="100%" height='100%' valign='bottom'>       <table border='0' cellpadding=0
cellspacing=0 height='100%' width='100%'>
            <tr>
              <td background='images/main_top_center.jpg' height='16' width='16' colspan='2'></
td>
            </tr>
            <tr>
              <td height='100%' width='100%' colspan='2'><iframe width='100%' height='100%'
scrolling='no' frameborder='0' name="main1"></iframe>
              </td>
            </tr>
            <tr>
```

```
            <td background='images/main_bottom_center.jpg' height='24' valign='bottom' width='100%'></td>
            <td background='images/main_bottom_center_right.jpg' height='24' width='300' valign='bottom'></td>
          </tr>
        </table></TD>
        <TD width="16" height='100%'>
          <TABLE border=0 cellPadding=0 cellSpacing=0 height='100%'>
            <TR>
              <TD background='images/main_right_top.jpg' height='16' width='16'>
              </TD>
            </TR>
            <TR>
              <TD background='images/main_right_middle.jpg' height='100%'>
              </TD>
            </TR>
            <TR>
              <TD background='images/main_right_bottom.jpg' height='30'>
              </TD>
            </TR>
          </TABLE>
        </TD>
      </tr>
      <tr>
        <td height="20" colspan="6" align='right' background='images/bodybackground.jpg'><font style='font-size:9pt'>第一开发小组   </font></td>
      </tr>
    </table>
  </body>
</html>


<%@ page language="java" import="java.util.*" pageEncoding="gb2312"%>
<%
String path = request.getContextPath();
String basePath = request.getScheme()+"://"+request.getServerName()+":"+request.getServerPort()+path+"/";
%>

<!DOCTYPE HTML PUBLIC "-//W3C//DTD HTML 4.01 Transitional//EN">
<html>
  <head>
    <base href="<%=basePath%>">
```

```html
<title> My JSP 'index_tree1.jsp' starting page </title>

<meta http-equiv="pragma" content="no-cache">
<meta http-equiv="cache-control" content="no-cache">
<meta http-equiv="expires" content="0">
<meta http-equiv="keywords" content="keyword1,keyword2,keyword3">
<meta http-equiv="description" content="This is my page">
<LINK href="css/css.css" type=text/css rel=stylesheet>

</head>

<body MS_POSITIONING="GridLayout" leftmargin="0" topmargin="0" marginwidth="0" marginheight="0" bgcolor="#E8E8E8" class='body'>
        <TABLE style="BORDER-COLLAPSE: collapse" cellSpacing=0 cellPadding=0 width="300" border='0'>
            <TR>
                <TD CLOSPAN='2' height='10'></TD>
            </Tr>
            <TR>
                <TD width=5></TD>
                <TD width=295>

<IMG height=22 src="images/rootmiddle.jpg" width=22 align=absMiddle border=0>
    <A href="javascript:onChange(2)">
        基础数据设置

    </A>
    <BR>
<SPAN id=child2 style="DISPLAY: 0" valign='middle'>
    <IMG height=22 src="images/line_tri.jpg" width=22 align=absMiddle border='0'> <IMG height=22 src="images/leaf.jpg" width=10 align=absMiddle border='0'>
            <A href="updatenotice.asp" target="main1">
        新设申请书

    </A>
    <BR>
        <SPAN id=child2 style="DISPLAY: 0" valign='middle'>
            <IMG height=22 src="images/line_tri.jpg" width=22 align=absMiddle border='0'> <IMG height=22 src="images/leaf.jpg" width=10 align=absMiddle border='0'>
                <A href="servlet/CantonListServlet" target="main1">
        行政区划
```

```
                </A>
                <BR>
            <SPAN id=child2 style="DISPLAY: 0" valign='middle'>
                <IMG height=22 src="images/line_tri.jpg" width=22 align=absMiddle border='0'> <IMG height=22 src="images/leaf.jpg" width=10 align=absMiddle border='0'>
                    <A href="updatenotice.asp" target="main1">
                        设备仪器清单
                    </A>
                    <BR>

                <IMG height=22 src="images/roottop.jpg" width=22 align=absMiddle border=0>
                    <A href="javascript:onChange(1)">
                        部级资质新设申请
                    </A>
                    <BR>

            </SPAN>

                <IMG height=22 src="images/rootmiddle.jpg" width=22 align=absMiddle border=0>
                    <A href="javascript:onChange(3)">
                        数据导入导出
                    </A>
                    <BR>

                <IMG height=22 src="images/rootmiddle.jpg" width=22 align=absMiddle border=0>
                    <A href="javascript:leafF('Login.jsp')">
                        退出
                    </A>
                    <BR>

            </TD></TR></TABLE>
    </body>
</html>


<%@ page language="java" import="java.util.*" pageEncoding="gb2312"%>
<%@ taglib uri="http://java.sun.com/jsp/jstl/core" prefix="c" %>
<%
String path = request.getContextPath();
String basePath = request.getScheme()+"://"+request.getServerName()+":"+request.getServerPort()+path+"/";
```

任务六：建设用地审批电子报盘管理系统行政区划模块

```
%>

<!DOCTYPE HTML PUBLIC "-//W3C//DTD HTML 4.01 Transitional//EN">
<html>
  <head>
    <base href="<%=basePath%>">

    <title>无标题文档</title>

    <meta http-equiv="pragma" content="no-cache">
    <meta http-equiv="cache-control" content="no-cache">
    <meta http-equiv="expires" content="0">
    <meta http-equiv="keywords" content="keyword1,keyword2,keyword3">
    <meta http-equiv="description" content="This is my page">
    <!--
    <link rel="stylesheet" type="text/css" href="styles.css">
    -->

  </head>
  <script type="text/javascript">
    function onAdd(){
        window.location="../addCanton.jsp";
    }
  </script>
  <body>
    <table width="764">
      <tr>
        <td align="right" colspan="5">
          <button type="button" onClick="onAdd()">添加行政区划</button>
        </td>
      </tr>
    </table><br/>
<c:choose>
<c:when test="${empty list}">
    <center><p>没有行政区划信息</p></center>
</c:when>
<c:otherwise>
<table width="764" border="1" style="border-collapse:collapse">
  <tr>
    <td width="83"><div align="center">行政区代码</div></td>
    <td width="107"><div align="center">行政区名称</div></td>
    <td width="125"><div align="center">政府名称</div></td>
    <td width="145"><div align="center">国土部门名称</div></td>
```

```
        <td width="144"><div align="center">上级行政区代码</div></td>
      </tr>
      <c:forEach var="Canton" items="${list}">
      <tr>
        <td><div align="center">${Canton.ctn_code}</div></td>
        <td><div align="center"><SPAN class=content>${Canton.ctn_name}</SPAN></div></td>
        <td><div align="center">${Canton.gov_name}</div></td>
        <td><div align="center">${Canton.land_dp_name}</div></td>
        <td><div align="center">${Canton.parent_code}</div></td>
      </tr>
      </c:forEach>
    </table>
  </c:otherwise>
  </c:choose>
</body>
</html>


    <%@ page language="java" import="java.util.*" pageEncoding="gb2312"%>
    <%
String path = request.getContextPath();
String basePath = request.getScheme()+"://"+request.getServerName()+":"+request.getServerPort()+path+"/";
    %>

    <!DOCTYPE HTML PUBLIC "-//W3C//DTD HTML 4.01 Transitional//EN">
    <html>
      <head>
        <base href="<%=basePath%>">

        <title>My JSP 'current_site.jsp' starting page</title>

        <meta http-equiv="pragma" content="no-cache">
        <meta http-equiv="cache-control" content="no-cache">
        <meta http-equiv="expires" content="0">
        <meta http-equiv="keywords" content="keyword1,keyword2,keyword3">
        <meta http-equiv="description" content="This is my page">
        <link href="css/style.css" rel=stylesheet type=text/css>

      </head>

      <body topmargin='0' leftmargin='0'>
        <table cellpadding="0" cellspacing="0" height='23' width='100%' border='0'>
```

```
        <tr>
            <td style="background-image: url('images/main_header_left.jpg'); background-repeat: no-repeat" height='23' width='32'>  </td>
            <td class='lth' valign='bottom' background='images/main_header_title.jpg' nowrap>当前位置:操作员 &gt;<font color="#FF0000">登录</font></td>
            <td background='images/main_header_center.jpg' height='23' width='100'>          </td>
            <td background='images/main_header_center1.jpg' height='23' width='100%'></td>
            <td background='images/main_header_right.jpg' height='23' width='30' align='right'>   </td>
        </tr>
    </table>
</body>
</html>


<%@ page language="java" import="java.util.*" pageEncoding="gb2312"%>
<%
String path = request.getContextPath();
String basePath = request.getScheme()+"://"+request.getServerName()+":"+request.getServerPort()+path+"/";
%>

<!DOCTYPE HTML PUBLIC "-//W3C//DTD HTML 4.01 Transitional//EN">
<html>
    <head>
        <base href="<%=basePath%>">

        <title>添加行政区划</title>

        <meta http-equiv="pragma" content="no-cache">
        <meta http-equiv="cache-control" content="no-cache">
        <meta http-equiv="expires" content="0">
        <meta http-equiv="keywords" content="keyword1,keyword2,keyword3">
        <meta http-equiv="description" content="This is my page">
        <!--
        <link rel="stylesheet" type="text/css" href="styles.css">
        -->

        <style type="text/css">
<!--
.style1 {color: #FF0000}
-->
        </style>
```

```
</head>

<body>
<script language = "javascript" >
    function check( ) {
        if ( form1. ctn_code. value = = "" ) {
            alert( "请输入行政区代码" )
            form1. ctn_code. focus( );
            return false;
        }
        if ( form1. ctn_name. value = = "" ) {
            alert( "请输入行政区名称" );
            form1. ctn_name. focus( );
            return false;
        }
        if ( form1. gov_name. value = = "" ) {
            alert( "请输入政府名称" )
            form1. gov_name. focus( );
            return false;
        }
        if ( form1. land_dp_name. value = = "" ) {
            alert( "请输入国土部门名称" );
            form1. land_dp_name. focus( );
            return false;
        }
        if ( form1. parent_code. value = = "" ) {
            alert( "请输入上级行政区代码" )
            form1. parent_code. focus( );
            return false;
        }
        return true;
    }
</script>
    < table width = "559" border = "0" >
< tr > < td width = "553" > < form action = "servlet/CantonAddServlet" method = "post" name = "form1" onSubmit = "return check( )" >
    < table width = "303" height = "136" border = "1" align = "center" style = "border – collapse:collapse" >
        < tr >
            < td height = "9" colspan = "2" > < span class = "titletxt" >行政区划信息录入(以下带 < span class = "style1" > * < /span >为必填项) < /span > < /td >
        < /tr >
        < tr >
```

```html
            <td width="113" height="9">行政区代码</td>
            <td width="174"><input name="ctn_code" type="text" id="ctn_code" size="20">
                <span class="style1">*</span></td>
          </tr>
          <tr>
            <td width="113" height="4">行政区名称</td>
            <td><input name="ctn_name" type="text" id="ctn_name" size="20">
                <span class="style1">*</span></td>
          </tr>
          <tr>
            <td width="113" height="20">政府名称</td>
            <td><input name="gov_name" type="text" id="gov_name" size="20"></td>
          </tr>
          <tr>
            <td width="113" height="20">国土部门名称</td>
            <td><input name="land_dp_name" type="text" id="land_dp_name" size="20"></td>
          </tr>
          <tr>
            <td width="113" height="20">上级行政区代码</td>
            <td><input name="parent_code" type="text" id="parent_code" size="20"></td>
          </tr>
          <tr>
            <td colspan="2" align="center"><button type="submit">添加</button></td>
          </tr>
        </table>
    </form></td>
  </tr>
</table>
</body>
</html>
```

3. 编写模型层代码

```java
package com.entity;

/**
 * 行政区划信息
 * @author 唐伟
 *
 */
public class Canton {
    private int ctn_code;//行政区代码
    private String ctn_name;//行政区名称
    private String gov_name;//政府名称
```

```java
        private String land_dp_name;//国土部门名称
        private int parent_code;//上级行政区代码

        public int getCtn_code() {
            return ctn_code;
        }
        public void setCtn_code(int ctn_code) {
            this.ctn_code = ctn_code;
        }
        public String getCtn_name() {
            return ctn_name;
        }
        public void setCtn_name(String ctn_name) {
            this.ctn_name = ctn_name;
        }
        public String getGov_name() {
            return gov_name;
        }
        public void setGov_name(String gov_name) {
            this.gov_name = gov_name;
        }
        public String getLand_dp_name() {
            return land_dp_name;
        }
        public void setLand_dp_name(String land_dp_name) {
            this.land_dp_name = land_dp_name;
        }
        public int getParent_code() {
            return parent_code;
        }
        public void setParent_code(int parent_code) {
            this.parent_code = parent_code;
        }

}
```

4. 编写控制层代码

```java
        package com.filter;

        import java.io.IOException;
        import javax.servlet.Filter;
        import javax.servlet.FilterChain;
        import javax.servlet.FilterConfig;
```

```java
import javax.servlet.ServletException;
import javax.servlet.ServletRequest;
import javax.servlet.ServletResponse;

/**
 *字符编码过滤器
 *@author 唐伟
 *
 */
public class CharSetFilter implements Filter{
    public void init(FilterConfig config) throws ServletException {

    }
    //执行 Filter
    public void doFilter(ServletRequest request, ServletResponse response,
        FilterChain chain) throws IOException, ServletException {

        request.setCharacterEncoding("gb2312");
        response.setCharacterEncoding("gb2312");
        chain.doFilter(request,response);
    }
    //释放
    public void destroy() {

    }

}


package com.servlet;

import java.io.IOException;
import java.io.PrintWriter;
import java.util.List;

import javax.servlet.ServletException;
import javax.servlet.http.HttpServlet;
import javax.servlet.http.HttpServletRequest;
import javax.servlet.http.HttpServletResponse;

import com.entity.Canton;
import com.service.CantonService;
import com.service.CantonServiceImpl;
```

```java
/**
 * 行政区划查询所有控制类
 * @author 唐伟
 *
 */
public class CantonListServlet extends HttpServlet {
    private CantonService cantonService;
    /**
     * The doGet method of the servlet. <br>
     *
     * This method is called when a form has its tag value method equals to get.
     *
     * @param request the request send by the client to the server
     * @param response the response send by the server to the client
     * @throws ServletException if an error occurred
     * @throws IOException if an error occurred
     */
    public void doGet(HttpServletRequest request, HttpServletResponse response)
        throws ServletException, IOException {

        this.doPost(request, response);
    }

    /**
     * The doPost method of the servlet. <br>
     *
     * This method is called when a form has its tag value method equals to post.
     *
     * @param request the request send by the client to the server
     * @param response the response send by the server to the client
     * @throws ServletException if an error occurred
     * @throws IOException if an error occurred
     */
    public void doPost(HttpServletRequest request, HttpServletResponse response)
            throws ServletException, IOException {

        response.setContentType("text/html");
        PrintWriter out = response.getWriter();
        cantonService = new CantonServiceImpl();
        List<Canton> list = cantonService.listCanton();
        request.setAttribute("list", list);
        request.getRequestDispatcher("../listCanton.jsp").forward(request, response);
        out.flush();
```

```
        out.close();
    }

}


package com.servlet;

import java.io.IOException;
import java.io.PrintWriter;

import javax.servlet.ServletException;
import javax.servlet.http.HttpServlet;
import javax.servlet.http.HttpServletRequest;
import javax.servlet.http.HttpServletResponse;

import com.entity.Canton;
import com.service.CantonService;
import com.service.CantonServiceImpl;

/**
 * 行政区划添加控制类
 * @author 唐伟
 *
 */
public class CantonAddServlet extends HttpServlet {
    private CantonService cantonService;
    /**
     * The doGet method of the servlet. <br>
     *
     * This method is called when a form has its tag value method equals to get.
     *
     * @param request the request send by the client to the server
     * @param response the response send by the server to the client
     * @throws ServletException if an error occurred
     * @throws IOException if an error occurred
     */
    public void doGet(HttpServletRequest request, HttpServletResponse response)
            throws ServletException, IOException {

        this.doPost(request, response);
    }

    /**
```

* The doPost method of the servlet.
 *
 * This method is called when a form has its tag value method equals to post.
 *
 * @param request the request send by the client to the server
 * @param response the response send by the server to the client
 * @throws ServletException if an error occurred
 * @throws IOException if an error occurred
 */
public void doPost(HttpServletRequest request, HttpServletResponse response)
        throws ServletException, IOException {

    response.setContentType("text/html");
    PrintWriter out = response.getWriter();
    int ctn_code = Integer.parseInt(request.getParameter("ctn_code"));
    String ctn_name = request.getParameter("ctn_name");
    String gov_name = request.getParameter("gov_name");
    String land_dp_name = request.getParameter("land_dp_name");
    int parent_code = Integer.parseInt(request.getParameter("parent_code"));
    Canton canton = new Canton();
    canton.setCtn_code(ctn_code);
    canton.setCtn_name(ctn_name);
    canton.setGov_name(gov_name);
    canton.setLand_dp_name(land_dp_name);
    canton.setParent_code(parent_code);
    cantonService = new CantonServiceImpl();
    if(cantonService.addCanton(canton)) {
        request.getRequestDispatcher("CantonListServlet").forward(request, response);
    } else {
        System.out.println("fail");
        request.getRequestDispatcher("../index.jsp");
    }
    out.flush();
    out.close();
}

}

package com.service;

import java.util.List;

import com.entity.Canton;

```java
/**
 *行政区划服务接口
 *@ author 唐伟
 *
 */
public interface CantonService {
    /**
     *添加行政区划信息
     *@ param canton
     *@ return
     */
    boolean addCanton(Canton canton);

    /**
     *修改行政区划信息
     *@ param canton
     *@ return
     */
    boolean updateCanton(Canton canton);

    /**
     *删除行政区划信息
     *@ param id
     *@ return
     */
    boolean deleteCanton(int id);

    /**
     *查询所有的行政区划信息
     *@ return
     */
    List<Canton> listCanton();

    /**
     *查询行政区划信息
     *@ param id
     *@ return
     */
    Canton findCanton(int id);
}

package com.service;
```

```java
import java.util.List;

import com.dao.CantonDao;
import com.dao.CantonDaoImpl;
import com.entity.Canton;

/**
 * 行政区划服务实现类
 * @author 唐伟
 *
 */
public class CantonServiceImpl implements CantonService {
    private CantonDao cantonDao;

    public CantonServiceImpl() {
        cantonDao = new CantonDaoImpl();
    }

    public boolean addCanton(Canton canton) {
        return cantonDao.addCanton(canton);
    }

    public boolean deleteCanton(int id) {
        // TODO Auto-generated method stub
        return false;
    }

    public Canton findCanton(int id) {
        // TODO Auto-generated method stub
        return null;
    }

    public List<Canton> listCanton() {
        return cantonDao.listCanton();
    }

    public boolean updateCanton(Canton canton) {
        // TODO Auto-generated method stub
        return false;
    }
}
```

任务七：建设用地审批电子报盘管理系统补偿标准模块

一、任务描述

你作为《建设用地审批电子报盘管理系统》项目开发组的程序员，请实现如下功能：
➢ 补偿标准信息的列表显示；
➢ 补偿标准信息的添加。

二、功能描述

（1）点击图7.1左边导航条中的"补偿标准"菜单项，则在右边的主体部分显示补偿标准信息列表。

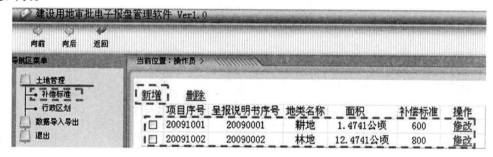

图7.1　补偿标准信息列表页面

（2）点击图7.2中的"新增"链接，则进入补偿标准信息录入页面，如图7.2所示。

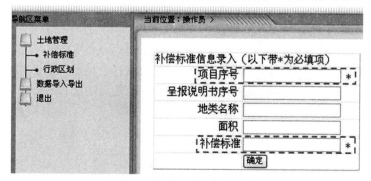

图7.2　补偿标准信息录入页面

(3)对图7.2中打"*"号的输入部分进行必填校验。

(4)点击图7.2中"确定"按钮,在补偿标准表中增加一条补偿标准信息。

(5)补偿标准信息增加成功后,自动定位到补偿标准信息列表页面,显示更新后的补偿标准信息列表,如图7.1所示。

(6)测试程序,通过补偿标准信息录入页面增加两条以上补偿标准信息。

三、要求

1. 界面实现

以提供的素材为基础,实现图7.1、图7.2所示页面。

2. 数据库实现

(1)创建数据库 LandDB。

(2)创建补偿标准信息表(T_requisiton_no_tilth),表结构见表7.1。

表7.1 补偿标准信息表(T_requisiton_no_tilth)表结构

字段名	字段说明	字段类型	允许为空	备注
Td_guid	项目序号	varchar(38)	否	主键
Bpl_guid	呈报说明书序号	varchar(38)	否	
Dl_name	地类名称	varchar(50)	是	
Area	面积	numeric(18,4)	是	数值型,单位为公顷
Std	补偿标准	numeric(18,4)	是	数值型,单位为万元

(3)在表 T_requisiton_no_tilth 插入记录,见表7.2。

表7.2 补偿标准信息表(T_requisiton_no_tilth)记录

Td_guid	Bpl_guid	Dl_name	Area	Std
20091001	20090001	耕地	1.4147	600
20091002	20090002	林地	12.4147	800

3. 功能实现

(1)功能需求如图7.3所示。

(2)依据补偿标准信息列表活动图完成补偿标准信息列表显示功能,如图7.4所示。

(3)依据添加补偿标准信息活动图完成添加补偿标准信息功能,如图7.5所示。

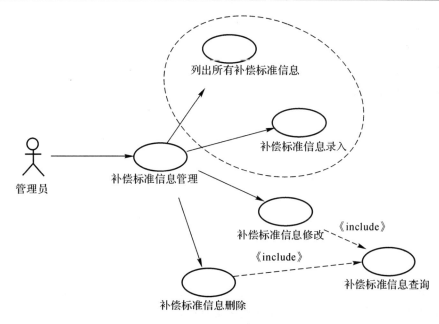

图 7.3 补偿标准信息设置模块用例图

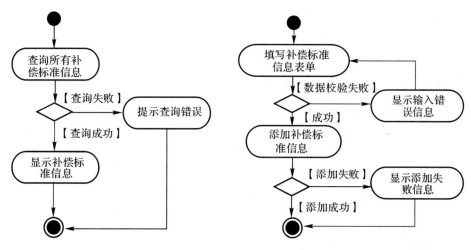

图 7.4 补偿标准信息列表活动图　　图 7.5 添加补偿标准信息活动图

四、必备知识

1. 数据库相关知识

（1）使用 MS SQL Server 2005/2008 创建数据库，创建数据表，设置表的字段，数据类型，主键，外键，约束。

（2）向数据表插入、删除、修改、查询数据。

2. 页面相关知识

（1）使用 HTML 制作项目页面。

（2）使用 CSS 控制页面的样式。

(3)使用JavaScript对页面必要的内容进行校验。

3. JSP相关知识

(1)使用JSTL标准标签库控制页面显示逻辑。
(2)理解JSP的request、response、session、application的概念。
(3)使用EL表达式在页面显示数据。
(4)合理地使用转发和重定向控制项目的页面跳转。
(5)使用JDBC与数据库进行交互。
(6)模式下的分层架构,控制器,视图,模型的划分和通信。

五、解题思路

1. 数据库思路

(1)根据项目要求创建数据库和数据表,向数据表中插入合适的测试数据。
(2)导入JDBC驱动包,编写JDBC的连接工具代码。
(3)编写数据操作对象代码,负责进行与数据库的交互操作。

2. 视图层思路

(1)将提供的素材页面改写为JSP页面。
(2)JSP使用JSTL和EL负责控制页面显示的逻辑。

3. 控制层思路

(1)使用Servlet类控制一次请求响应过程的处理。
(2)由Servlet按照顺序进行请求的处理、数据库交互、模型存取和封装、页面跳转逻辑控制等。

4. 模型层思路

使用JavaBean作为模型层,封装数据和行为。

六、操作步骤

1. 准备数据库

(1)根据项目要求,在SQL Server2008中创建LandDB数据库、补偿标准信息表(T_requisiton_no_tilth),并插入测试数据。
(2)编写JDBC的连接工具代码。

```
package com.dao;

import java.sql.Connection;
import java.sql.DriverManager;
import java.sql.PreparedStatement;
import java.sql.ResultSet;
```

```java
import java.sql.SQLException;

/**
 * 数据库连接类
 * @author 唐伟
 *
 */
public class DBConnection {
    private static final String DRIVER = "com.microsoft.sqlserver.jdbc.SQLServerDriver";
    private static final String URL = "jdbc:sqlserver://localhost:1433;DatabaseName=LandDB";
    private static final String USER = "sa";
    private static final String PWD = "123";

    private static Connection con;
    private static PreparedStatement prst;
    private static ResultSet rest;

    /**
     * 加载驱动
     */
    static {
        try {
            Class.forName(DRIVER);
        } catch (ClassNotFoundException e) {
            e.printStackTrace();
        }
    }

    /**
     * 取得连接
     * @return
     */
    public static Connection getConnection() {
        try {
            con = DriverManager.getConnection(URL, USER, PWD);
        } catch (SQLException e) {
            e.printStackTrace();
        }
        return con;
    }

    /**
     * 关闭连接
```

```java
         */
        public static void close() {
            try {
                if(rest! = null) {
                    rest.close();
                }
                if(prst! = null) {
                    prst.close();
                }
                if(con! = null) {
                    con.close();
                }
            } catch (SQLException e) {
                e.printStackTrace();
            }
        }
    }


package com.dao;

import java.util.List;

import com.entity.Requisition;

/**
 * 补偿标准操作接口
 * @author 唐伟
 *
 */
public interface RequisitionDao {
    /**
     * 添加补偿标准信息
     * @param canton
     * @return
     */
    boolean addCanton(Requisition canton);

    /**
     * 修改补偿标准信息
     * @param canton
     * @return
     */
    boolean updateCanton(Requisition canton);
```

```
/**
 * 删除补偿标准信息
 * @param id
 * @return
 */
boolean deleteCanton(int id);

/**
 * 查询所有的补偿标准信息
 * @return
 */
List<Requisition> listCanton();

/**
 * 查询补偿标准信息
 * @param id
 * @return
 */
Requisition findCanton(int id);
}
```

```
package com.dao;

import java.sql.Connection;
import java.sql.PreparedStatement;
import java.sql.ResultSet;
import java.sql.SQLException;
import java.util.ArrayList;
import java.util.List;

import com.entity.Requisition;

/**
 * 补偿标准操作实现类
 * @author 唐伟
 *
 */
public class RequisitionDaoImpl implements RequisitionDao {

    public boolean addCanton(Requisition canton) {
        boolean flag = false;
        String sql = "insert into T_REQUISITION_NO_TILTH values(?,?,?,?,?)";
```

```java
        Connection con = null;
        PreparedStatement prst = null;
        con = DBConnection.getConnection();
        try {
            prst = con.prepareStatement(sql);
            prst.setString(1, canton.getTd_guid());
            prst.setString(2, canton.getBpl_guid());
            prst.setString(3, canton.getDl_name());
            prst.setDouble(4, canton.getArea());
            prst.setString(5, canton.getStd());
            if(prst.executeUpdate() != 0) {
                flag = true;
            }
        } catch (SQLException e) {
            e.printStackTrace();
        } finally {
            DBConnection.close();
        }
        return flag;
    }

    public boolean deleteCanton(int id) {
        // TODO Auto-generated method stub
        return false;
    }

    public Requisition findCanton(int id) {
        // TODO Auto-generated method stub
        return null;
    }

    public List<Requisition> listCanton() {
        List<Requisition> list = new ArrayList<Requisition>();
        String sql = "select * from T_REQUISITION_NO_TILTH";
        Connection con = null;
        PreparedStatement prst = null;
        ResultSet rest = null;
        con = DBConnection.getConnection();
        try {
            prst = con.prepareStatement(sql);
            rest = prst.executeQuery();
            while(rest.next()) {
                Requisition canton = new Requisition();
```

```
                canton.setTd_guid(rest.getString("td_guid"));
                canton.setBpl_guid(rest.getString("bpl_guid"));
                canton.setDl_name(rest.getString("dl_name"));
                canton.setArea(rest.getDouble("area"));
                canton.setStd(rest.getString("std"));
                list.add(canton);
            }
        } catch (SQLException e) {
            e.printStackTrace();
        }
        return list;
    }

    public boolean updateCanton(Requisition canton) {
        // TODO Auto-generated method stub
        return false;
    }

}
```

2. 编写视图层代码

```
<%@ page language="java" import="java.util.*" pageEncoding="gb2312"%>
<%@ taglib uri="http://java.sun.com/jsp/jstl/core" prefix="c" %>
<%
String path = request.getContextPath();
String basePath = request.getScheme()+"://"+request.getServerName()+":"+request.getServerPort()+path+"/";
%>

<!DOCTYPE HTML PUBLIC "-//W3C//DTD HTML 4.01 Transitional//EN">
<html>
  <head>
    <base href="<%=basePath%>">

    <title>My JSP 'listRequisition.jsp' starting page</title>

    <meta http-equiv="pragma" content="no-cache">
    <meta http-equiv="cache-control" content="no-cache">
    <meta http-equiv="expires" content="0">
    <meta http-equiv="keywords" content="keyword1,keyword2,keyword3">
    <meta http-equiv="description" content="This is my page">
    <!--
```

```html
<link rel="stylesheet" type="text/css" href="styles.css">
-->

</head>

<body>
  <form name="form1" method="post" action="">
<table width="627">
<tr>
    <td width="54"><div align="left"><a href="addRequisition.jsp">新增</a></div></td>
    <td width="46"><div align="right"><a href="#">删除</a></div></td>
    <td width="511"> </td>
</tr>
<c:choose>
<c:when test="${empty list}">
    <center><p>没有补偿标准信息</p></center>
</c:when>
<c:otherwise>
  <tr>
    <td colspan="3"><table width="566" border="1" style="border-collapse:collapse">
    <tr>
        <td width="32"><div align="center"></div></td>
        <td width="70"><div align="center">项目序号</div></td>
        <td width="120"><div align="center">呈报说明书序号</div></td>
        <td width="70"><div align="center">地类名称</div></td>
        <td width="87"><div align="center">面积</div></td>
        <td width="68"><div align="center">补偿标准</div></td><td width="73"><div align="center">操作</div></td>
    </tr>
    <c:forEach var="requisition" items="${list}">
    <tr>
    <td><div align="center">
        <input type="checkbox" name="checkbox" value="checkbox">
    </div></td>
    <td><div align="center">${requisition.td_guid}</div></td>
    <td><div align="center">${requisition.bpl_guid}</div></td>
    <td><div align="center">${requisition.dl_name}<BR>
    </div></td>
    <td><div align="center">${requisition.area}公顷</div></td>
```

```
            <td><div align="center">${requisition.std}</div></td>
            <td><div align="center"><a href="addstaff.htm">修改</a></div></td>
          </tr>
         </c:forEach>
       </table></td>
      </tr>
     </c:otherwise>
    </c:choose>
   </table>
  </form>
 </body>
</html>
```

```
<%@ page language="java" import="java.util.*" pageEncoding="gb2312"%>
<%
String path = request.getContextPath();
String basePath = request.getScheme()+"://"+request.getServerName()+":"+request.getServerPort()+path+"/";
%>

<!DOCTYPE HTML PUBLIC "-//W3C//DTD HTML 4.01 Transitional//EN">
<html>
  <head>
    <base href="<%=basePath%>">

    <title>My JSP 'addRequisition.jsp' starting page</title>

    <meta http-equiv="pragma" content="no-cache">
    <meta http-equiv="cache-control" content="no-cache">
    <meta http-equiv="expires" content="0">
    <meta http-equiv="keywords" content="keyword1,keyword2,keyword3">
    <meta http-equiv="description" content="This is my page">
    <!--
    <link rel="stylesheet" type="text/css" href="styles.css">
    -->

  </head>

  <body>
    <script language="javascript">
function check(){
   if(form1.td_guid.value==""){
      alert("请输入项目序号")
```

```
            form1.td_guid.focus();
            return false;
        }
        if(form1.bpl_guid.value==""){
            alert("请输入呈报说明书序号");
            form1.bpl_guid.focus();
            return false;
        }
    }
</script>
<table width="377" border="0">
<tr><td width="371"><form action="servlet/RequisitionAddServlet" method="post" name="form1" onSubmit="return check()">
    <table width="346">
        <tr>
            <td width="330" colspan="3"><table width="313" height="109" border="1" align="center" style="border-collapse:collapse">
                <tr>
                    <td width="131" height="9"><div align="right"><font color="red">*</font>项目序号</div></td>
                    <td width="166"><input name="td_guid" type="text" id="td_guid" size="20"></td>
                </tr>
                <tr>
                    <td width="131" height="4"><div align="right"><font color="red">*</font>呈报说明书序号</div></td>
                    <td><input name="bpl_guid" type="text" id="bpl_guid" size="20"></td>
                </tr>
                <tr>
                    <td width="131" height="20"><div align="right">地类名称</div></td>
                    <td><input name="dl_name" type="text" id="dl_name" size="20"></td>
                </tr>
                <tr>
                    <td width="131" height="20"><div align="right">面积</div></td>
                    <td><input name="area" type="text" id="area" size="20"></td>
                </tr>
                <tr>
                    <td height="20"><div align="right">补偿标准</div></td>
                    <td><input name="std" type="text" id="std" size="20"></td>
                </tr>
                <tr>
                    <td width="131" height="20"><div align="right"></div></td>
                    <td><input name="bt1" type="submit" id="bt1" value="确定"></td>
```

```
            </tr>
          </table> </td>
      </tr>
    </table>
  </form> </td>
</tr>
</table>
</body>
</html>
```

3. 编写模型层代码

```
package com.entity;

/**
 * 补偿标准信息
 * @author 唐伟
 *
 */
public class Requisition {
    private String td_guid;//项目序号
    private String bpl_guid;//呈报说明书序号
    private String dl_name;//地类名称
    private double area;//面积
    private String std;//补偿标准

    public String getTd_guid() {
        return td_guid;
    }
    public void setTd_guid(String td_guid) {
        this.td_guid = td_guid;
    }
    public String getBpl_guid() {
        return bpl_guid;
    }
    public void setBpl_guid(String bpl_guid) {
        this.bpl_guid = bpl_guid;
    }
    public String getDl_name() {
        return dl_name;
    }
    public void setDl_name(String dl_name) {
        this.dl_name = dl_name;
    }
```

```java
        public double getArea() {
            return area;
        }
        public void setArea(double area) {
            this.area = area;
        }
        public String getStd() {
            return std;
        }
        public void setStd(String std) {
            this.std = std;
        }
}
```

4. 编写控制层代码

```java
package com.servlet;

import java.io.IOException;
import java.io.PrintWriter;
import java.util.List;

import javax.servlet.ServletException;
import javax.servlet.http.HttpServlet;
import javax.servlet.http.HttpServletRequest;
import javax.servlet.http.HttpServletResponse;

import com.entity.Requisition;
import com.service.RequisitionService;
import com.service.RequisitionServiceImpl;

/**
 * 补偿标准查询所有控制类
 * @author 唐伟
 *
 */
public class RequisitionListServlet extends HttpServlet {
    private RequisitionService cantonService;
    /**
     * The doGet method of the servlet. <br>
     *
     * This method is called when a form has its tag value method equals to get.
     *
     * @param request the request send by the client to the server
```

```java
 * @param response the response send by the server to the client
 * @throws ServletException if an error occurred
 * @throws IOException if an error occurred
 */
public void doGet(HttpServletRequest request, HttpServletResponse response)
        throws ServletException, IOException {

    this.doPost(request, response);

}


/**
 * The doPost method of the servlet. <br>
 *
 * This method is called when a form has its tag value method equals to post.
 *
 * @param request the request send by the client to the server
 * @param response the response send by the server to the client
 * @throws ServletException if an error occurred
 * @throws IOException if an error occurred
 */
public void doPost(HttpServletRequest request, HttpServletResponse response)
        throws ServletException, IOException {

    response.setContentType("text/html");
    PrintWriter out = response.getWriter();
    cantonService = new RequisitionServiceImpl();
    List<Requisition> list = cantonService.listCanton();
    request.setAttribute("list", list);
    request.getRequestDispatcher("../listRequisition.jsp").forward(request, response);
    out.flush();
    out.close();

}

}


package com.servlet;

import java.io.IOException;
import java.io.PrintWriter;

import javax.servlet.ServletException;
    import javax.servlet.http.HttpServlet;
    import javax.servlet.http.HttpServletRequest;
```

```java
import javax.servlet.http.HttpServletResponse;

import com.entity.Requisition;
import com.service.RequisitionService;
import com.service.RequisitionServiceImpl;

/**
 * 补偿标准添加控制类
 * @author 唐伟
 *
 */
public class RequisitionAddServlet extends HttpServlet {
    private RequisitionService cantonService;
    /**
     * The doGet method of the servlet. <br>
     *
     * This method is called when a form has its tag value method equals to get.
     *
     * @param request the request send by the client to the server
     * @param response the response send by the server to the client
     * @throws ServletException if an error occurred
     * @throws IOException if an error occurred
     */
    public void doGet(HttpServletRequest request, HttpServletResponse response)
            throws ServletException, IOException {

        this.doPost(request, response);
    }

    /**
     * The doPost method of the servlet. <br>
     *
     * This method is called when a form has its tag value method equals to post.
     *
     * @param request the request send by the client to the server
     * @param response the response send by the server to the client
     * @throws ServletException if an error occurred
     * @throws IOException if an error occurred
     */
    public void doPost(HttpServletRequest request, HttpServletResponse response)
            throws ServletException, IOException {

        response.setContentType("text/html");
```

```java
            PrintWriter out = response.getWriter();
            String td_guid = request.getParameter("td_guid");
            String bpl_guid = request.getParameter("bpl_guid");
            Requisition requisition = new Requisition();
            requisition.setTd_guid(td_guid);
            requisition.setBpl_guid(bpl_guid);
            if(request.getParameter("dl_name") != null){
                String dl_name = request.getParameter("dl_name");
                requisition.setDl_name(dl_name);
            } else {
                requisition.setDl_name("");
            }
            if(!"".equals(request.getParameter("area"))){
                double area = Double.parseDouble(request.getParameter("area"));
                requisition.setArea(area);
            } else {
                requisition.setArea(0.0);
            }
            if(request.getParameter("std") != null){
                String std = request.getParameter("std");
                requisition.setStd(std);
            } else {
                requisition.setStd("");
            }
            cantonService = new RequisitionServiceImpl();
            if(cantonService.addCanton(requisition)){
                request.getRequestDispatcher("RequisitionListServlet").forward(request, response);
            } else {
                System.out.println("fail");
                request.getRequestDispatcher("../index.jsp");
            }
            out.flush();
            out.close();
        }
}

package com.service;

import java.util.List;

import com.entity.Requisition;
```

```java
/**
 * 补偿标准服务接口
 * @author 唐伟
 *
 */
public interface RequisitionService {
    /**
     * 添加补偿标准信息
     * @param canton
     * @return
     */
    boolean addCanton(Requisition canton);

    /**
     * 修改补偿标准信息
     * @param canton
     * @return
     */
    boolean updateCanton(Requisition canton);

    /**
     * 删除补偿标准信息
     * @param id
     * @return
     */
    boolean deleteCanton(int id);

    /**
     * 查询所有的补偿标准信息
     * @return
     */
    List<Requisition> listCanton();

    /**
     * 查询补偿标准信息
     * @param id
     * @return
     */
    Requisition findCanton(int id);
}


package com.service;
```

```java
import java.util.List;

import com.dao.RequisitionDao;
import com.dao.RequisitionDaoImpl;
import com.entity.Requisition;

/**
 * 补偿标准服务实现类
 * @author 唐伟
 *
 */
public class RequisitionServiceImpl implements RequisitionService {
    private RequisitionDao cantonDao;

    public RequisitionServiceImpl() {
        cantonDao = new RequisitionDaoImpl();
    }

    public boolean addCanton(Requisition canton) {
        return cantonDao.addCanton(canton);
    }

    public boolean deleteCanton(int id) {
        // TODO Auto-generated method stub
        return false;
    }

    public Requisition findCanton(int id) {
        // TODO Auto-generated method stub
        return null;
    }

    public List<Requisition> listCanton() {
        return cantonDao.listCanton();
    }

    public boolean updateCanton(Requisition canton) {
        // TODO Auto-generated method stub
        return false;
    }

}
```

任务八：建设用地审批电子报盘管理系统审批模块

一、任务描述

你作为《建设用地审批电子报盘管理系统》项目开发组的程序员，请实现如下功能：
- 审批信息的列表显示；
- 审批信息的删除。

二、功能描述

（1）点击图8.1中左边导航条中的"审批信息"，则在右边的主体部分显示审批信息列表。

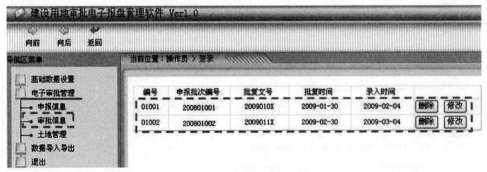

图8.1　审批信息列表页面

（2）点击图8.1中的"删除"按钮，则进入删除确认对话框，如图8.2所示。

图8.2　审批信息删除确认对话框页面

（3）点击图8.2中"确定"按钮，在审批信息表中删除一条审批信息。
（4）审批信息删除成功后，自动定位到审批信息列表页面，显示更新后的审批信息列表，如图8.1所示。

(5)测试程序,在审批页面中删除两条审批信息。

三、要求

1. 界面实现

以提供的素材为基础,实现图 8.1、图 8.2 所示页面。

2. 数据库实现

(1)创建数据库 LandDB。

(2)创建审批信息表(T_ministry_approve),表结构见表 8.1。

表 8.1　审批信息表(T_ministry_approve)表结构

字段名	字段说明	字段类型	允许为空	备注
Mi_guid	主键 ID	varchar(38)	否	主键
Proj_guid	申报批次编号	varchar(38)	否	
Approve_symbol	批复文号	varchar(20)	是	
Approve_time	批复时间	datetime	是	日期型
Submit_time	录入时间	datetime	是	日期型

(3)在表 T_ministry_approve 插入记录,见表 8.2。

表 8.2　审批信息表(T_ministry_approve)记录

Mi_guid	Proj_guid	Approve_symbol	Approve_time	Submit_time
01001	200801001	200902001	2009-1-30	2009-2-4
01002	200801002	200902002	2009-2-30	2009-3-4

3. 功能实现

(1)功能需求如图 8.3 所示。

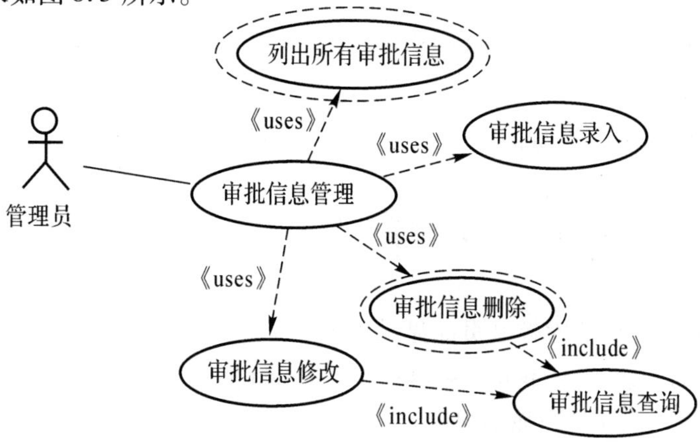

图 8.3　审批信息管理模块用例图

（2）依据审批信息列表活动图完成审批信息列表显示功能，如图8.4所示。
（3）依据删除审批信息活动图完成删除审批信息功能，如图8.5所示。

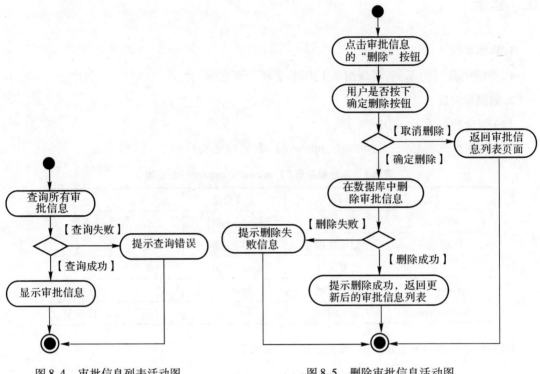

图8.4　审批信息列表活动图　　　　图8.5　删除审批信息活动图

四、必备知识

1. 数据库相关知识

（1）使用 MS SQL Server 2005/2008 创建数据库，创建数据表，设置表的字段，数据类型，主键，外键，约束。

（2）向数据表插入、删除、修改、查询数据。

2. 页面相关知识

（1）使用 HTML 制作项目页面。

（2）使用 CSS 控制页面的样式。

（3）使用 JavaScript 对页面必要的内容进行校验。

2. JSP 相关知识

（1）使用 JSTL 标准标签库控制页面显示逻辑。

（2）理解 JSP 的 request、response、session、application 的概念。

（3）使用 EL 表达式在页面显示数据。

（4）合理地使用转发和重定向控制项目的页面跳转。

（5）使用 JDBC 与数据库进行交互。

(6)MVC 模式下的分层架构,控制器,视图,模型的划分和通信。

五、解题思路

1. 数据库思路

(1)根据项目要求创建数据库和数据表,向数据表中插入合适的测试数据。
(2)导入 JDBC 驱动包,编写 JDBC 的连接工具代码。
(3)编写数据操作对象代码,负责进行与数据库的交互操作。

2. 视图层思路

(1)将提供的素材页面改写为 JSP 页面。
(2)JSP 使用 JSTL 和 EL 负责控制页面显示的逻辑。

3. 控制层思路

(1)使用 Servlet 类控制一次请求响应过程的处理。
(2)由 Servlet 按照顺序进行请求的处理、数据库交互、模型存取和封装、页面跳转逻辑控制等。

4. 模型层思路

使用 JavaBean 作为模型层,封装数据和行为。

六、操作步骤

1. 准备数据库

(1)根据项目要求,在 SQL Server2008 中创建 LandDB 数据库、审批信息表(T_ministry_approve),并插入测试数据。
(2)编写 JDBC 的连接工具代码。

```
package com.dao;

import java.sql.Connection;
import java.sql.DriverManager;
import java.sql.PreparedStatement;
import java.sql.ResultSet;
import java.sql.SQLException;

/**
 * 数据库连接类
 * @author 唐伟
 *
 */
public class DBConnection {
    private static final String DRIVER = "com.microsoft.sqlserver.jdbc.SQLServerDriver";
```

```java
    private static final String URL = "jdbc:sqlserver://localhost:1433;DatabaseName=LandDB";
    private static final String USER = "sa";
    private static final String PWD = "123";

    private static Connection con;
    private static PreparedStatement prst;
    private static ResultSet rest;

    /**
     * 加载驱动
     */
    static {
        try {
            Class.forName(DRIVER);
        } catch (ClassNotFoundException e) {
            e.printStackTrace();
        }
    }

    /**
     * 取得连接
     * @return
     */
    public static Connection getConnection() {
        try {
            con = DriverManager.getConnection(URL, USER, PWD);
        } catch (SQLException e) {
            e.printStackTrace();
        }
        return con;
    }

    /**
     * 关闭连接
     */
    public static void close() {
        try {
            if(rest != null) {
                rest.close();
            }
            if(prst != null) {
                prst.close();
```

```java
            if(con! = null){
                con.close();
            }
        } catch (SQLException e){
            e.printStackTrace();
        }
    }
}

package com.dao;

import java.util.List;

import com.entity.Approve;

/**
 * 审批信息操作接口
 * @author 唐伟
 *
 */
public interface ApproveDao{
    /**
     * 添加审批信息
     * @param canton
     * @return
     */
    boolean addApprove(Approve approve);

    /**
     * 修改审批信息
     * @param canton
     * @return
     */
    boolean updateApprove(Approve approve);

    /**
     * 删除审批信息
     * @param id
     * @return
     */
    boolean deleteApprove(String id);

    /**
```

```java
     * 查询所有的审批信息
     * @return
     */
    List<Approve> listApprove();

    /**
     * 查询审批信息
     * @param id
     * @return
     */
    Approve findApprove(int id);
}


package com.dao;

import java.sql.Connection;
import java.sql.PreparedStatement;
import java.sql.ResultSet;
import java.sql.SQLException;
import java.util.ArrayList;
import java.util.List;

import com.entity.Approve;

/**
 * 审批信息操作实现类
 * @author 唐伟
 *
 */
public class ApproveDaoImpl implements ApproveDao {

    public boolean addApprove(Approve approve) {
        return false;
    }

    public boolean deleteApprove(String id) {
        String sql = "delete from T_MINISTRY_APPROVE where MI_GUID = ?";
        Connection con = null;
        PreparedStatement prst = null;
        con = DBConnection.getConnection();
        try {
            prst = con.prepareStatement(sql);
            prst.setString(1, id);
```

```java
            if( prst. executeUpdate( )! =0) {
                return true;
            }
        } catch (SQLException e) {
            e. printStackTrace( );
        }
        return false;
    }

    public Approve findApprove( int id) {
        // TODO Auto-generated method stub
        return null;
    }

    public List < Approve > listApprove( ) {
        List < Approve > list = new ArrayList < Approve > ( );
        String sql = "select * from T_MINISTRY_APPROVE";
        Connection con = null;
        PreparedStatement prst = null;
        ResultSet rest = null;
        con = DBConnection. getConnection( );
        try {
            prst = con. prepareStatement( sql) ;
            rest = prst. executeQuery( ) ;
            while( rest. next( ) ) {
                Approve approve = new Approve( );
                approve. setMi_guid( rest. getString( "mi_guid" ) );
                approve. setProj_guid( rest. getString( "proj_guid" ) );
                approve. setApprove_symbol( rest. getString( "approve_symbol" ) );
                approve. setApprove_time( rest. getString( "approve_time" ) );
                approve. setSubmit_time( rest. getString( "submit_time" ) );
                list. add( approve) ;
            }
        } catch (SQLException e) {
            e. printStackTrace( );
        }
        return list;
    }

    public boolean updateApprove( Approve approve) {
        // TODO Auto-generated method stub
        return false;
    }
```

    }

2. 编写视图层代码

```jsp
<%@ page language="java" import="java.util.*" pageEncoding="gb2312"%>
<%@ taglib uri="http://java.sun.com/jsp/jstl/core" prefix="c" %>
<%
String path = request.getContextPath();
String basePath = request.getScheme()+"://"+request.getServerName()+":"+request.getServerPort()+path+"/";
%>

<!DOCTYPE HTML PUBLIC "-//W3C//DTD HTML 4.01 Transitional//EN">
<html>
  <head>
    <base href="<%=basePath%>">

    <title>My JSP 'listApprove.jsp' starting page</title>

    <meta http-equiv="pragma" content="no-cache">
    <meta http-equiv="cache-control" content="no-cache">
    <meta http-equiv="expires" content="0">
    <meta http-equiv="keywords" content="keyword1,keyword2,keyword3">
    <meta http-equiv="description" content="This is my page">
    <!--
    <link rel="stylesheet" type="text/css" href="styles.css">
    -->

  </head>
  <script language="javascript">
  function del(id){
      if(window.confirm("确认要删除此信息吗?")){
          window.location.href="ApproveDeleteServlet?id="+id;
          return true;
      }
      return false;
  }
  </script>
  <body>
    <c:choose>
      <c:when test="${empty list}">
        <center><p>没有审批信息</p></center>
      </c:when>
```

任务八：建设用地审批电子报盘管理系统审批模块

```html
<c:otherwise>
    <table width="573" border="1" style="border-collapse:collapse">
        <tr>
            <td width="54"><div align="center"><font color="#3074A2" style="font-size:9pt;color:#000000">编号</font></div></td>
            <td width="94"><div align="center"><font color="#3074A2" style="font-size:9pt;color:#000000">申报批次编号</font></div></td>
            <td width="90"><div align="center"><font color="#3074A2" style="font-size:9pt;color:#000000">批复文号</font></div></td>
            <td width="96"><div align="center"><font color="#3074A2" style="font-size:9pt;color:#000000">批复时间</font></div></td>
            <td width="95"><div align="center"><font color="#3074A2" style="font-size:9pt;color:#000000">录入时间</font></div>   <div align="center"></div>   </td>
            <td width="104"> </td>
        </tr>
        <c:forEach var="approve" items="${list}">
        <tr>
            <td><div align="center"><font color="#3074A2" style="font-size:9pt;color:#000000">${approve.mi_guid}</font></div></td>
            <td><div align="center"><font color="#3074A2" style="font-size:9pt;color:#000000">${approve.proj_guid}</font> </div></td>
            <td><div align="center"><font color="#3074A2" style="font-size:9pt;color:#000000">${approve.approve_symbol}</font></div></td>
            <td><div align="center"><font color="#3074A2" style="font-size:9pt;color:#000000">${approve.approve_time}</font></div></td>
            <td><div align="center"><font color="#3074A2" style="font-size:9pt;color:#000000"></font></div>
                <div align="center"><font color="#3074A2" style="font-size:9pt;color:#000000">${approve.submit_time}</font></div></td>
            <td>
                <input type="button" name="bt1" value="删除" onClick="return del('${approve.mi_guid}')">
                <input type="submit" name="bt2" value="修改">
            </td>
        </tr>
        </c:forEach>
    </table>
</c:otherwise>
</c:choose>
</body>
</html>
```

3. 编写模型层代码

```java
package com.entity;

/**
 * 审批信息
 * @author 唐伟
 *
 */
public class Approve {
    private String mi_guid;//主键ID
    private String proj_guid;//申报批次编号
    private String approve_symbol;//批复文号
    private String approve_time;//批复时间
    private String submit_time;//录入时间

    public String getMi_guid() {
        return mi_guid;
    }
    public void setMi_guid(String mi_guid) {
        this.mi_guid = mi_guid;
    }
    public String getProj_guid() {
        return proj_guid;
    }
    public void setProj_guid(String proj_guid) {
        this.proj_guid = proj_guid;
    }
    public String getApprove_symbol() {
        return approve_symbol;
    }
    public void setApprove_symbol(String approve_symbol) {
        this.approve_symbol = approve_symbol;
    }
    public String getApprove_time() {
        return approve_time;
    }
    public void setApprove_time(String approve_time) {
        this.approve_time = approve_time;
    }
    public String getSubmit_time() {
        return submit_time;
    }
```

```java
        public void setSubmit_time(String submit_time){
            this.submit_time = submit_time;
        }
    }
```

4. 编写控制层代码

```java
package com.servlet;

import java.io.IOException;
import java.io.PrintWriter;
import java.util.List;

import javax.servlet.ServletException;
import javax.servlet.http.HttpServlet;
import javax.servlet.http.HttpServletRequest;
import javax.servlet.http.HttpServletResponse;

import com.entity.Approve;
import com.service.ApproveService;
import com.service.ApproveServiceImpl;

/**
 * 审批信息查询所有控制类
 * @author 唐伟
 *
 */
public class ApproveListServlet extends HttpServlet {
    private ApproveService cantonService;
    /**
     * The doGet method of the servlet. <br>
     *
     * This method is called when a form has its tag value method equals to get.
     *
     * @param request the request send by the client to the server
     * @param response the response send by the server to the client
     * @throws ServletException if an error occurred
     * @throws IOException if an error occurred
     */
    public void doGet(HttpServletRequest request, HttpServletResponse response)
            throws ServletException, IOException {

        this.doPost(request, response);
    }
```

```
/**
 * The doPost method of the servlet. <br>
 *
 * This method is called when a form has its tag value method equals to post.
 *
 * @param request the request send by the client to the server
 * @param response the response send by the server to the client
 * @throws ServletException if an error occurred
 * @throws IOException if an error occurred
 */
public void doPost(HttpServletRequest request, HttpServletResponse response)
    throws ServletException, IOException {

    response.setContentType("text/html");
    PrintWriter out = response.getWriter();
    cantonService = new ApproveServiceImpl();
    List<Approve> list = cantonService.listApprove();
    request.setAttribute("list", list);
    request.getRequestDispatcher("../listApprove.jsp").forward(request, response);
    out.flush();
    out.close();
  }

}


package com.servlet;

import java.io.IOException;
import java.io.PrintWriter;

import javax.servlet.ServletException;
import javax.servlet.http.HttpServlet;
import javax.servlet.http.HttpServletRequest;
import javax.servlet.http.HttpServletResponse;

import com.entity.Approve;
import com.service.ApproveService;
import com.service.ApproveServiceImpl;

/**
 * 审批信息删除控制类
 * @author 唐伟
 *
```

```java
 */
public class ApproveDeleteServlet extends HttpServlet {
    private ApproveService cantonService;
    /**
     * The doGet method of the servlet. <br>
     *
     * This method is called when a form has its tag value method equals to get.
     *
     * @param request the request send by the client to the server
     * @param response the response send by the server to the client
     * @throws ServletException if an error occurred
     * @throws IOException if an error occurred
     */
    public void doGet(HttpServletRequest request, HttpServletResponse response)
            throws ServletException, IOException {

        this.doPost(request, response);
    }

    /**
     * The doPost method of the servlet. <br>
     *
     * This method is called when a form has its tag value method equals to post.
     *
     * @param request the request send by the client to the server
     * @param response the response send by the server to the client
     * @throws ServletException if an error occurred
     * @throws IOException if an error occurred
     */
    public void doPost(HttpServletRequest request, HttpServletResponse response)
            throws ServletException, IOException {

        response.setContentType("text/html");
        PrintWriter out = response.getWriter();
        String id = request.getParameter("id");
        System.out.println(id);
        cantonService = new ApproveServiceImpl();
        if(cantonService.deleteApprove(id)) {
            request.getRequestDispatcher("ApproveListServlet").forward(request, response);
        } else {
            System.out.println("fail");
            request.getRequestDispatcher("../error.html").forward(request, response);
        }
```

```java
            out.flush();
            out.close();
        }

}

package com.service;

import java.util.List;

import com.entity.Approve;

/**
 * 审批信息服务接口
 * @author 唐伟
 *
 */
public interface ApproveService {
    /**
     * 添加审批信息
     * @param canton
     * @return
     */
    boolean addApprove(Approve approve);

    /**
     * 修改审批信息
     * @param canton
     * @return
     */
    boolean updateApprove(Approve approve);

    /**
     * 删除审批信息
     * @param id
     * @return
     */
    boolean deleteApprove(String id);

    /**
     * 查询所有的审批信息
     * @return
     */
```

```java
    List < Approve >  listApprove( ) ;

    /**
     * 查询审批信息
     * @param id
     * @return
     */
    Approve findApprove( int id) ;
}


package com. service;

import java. util. List;

import com. dao. ApproveDao;
import com. dao. ApproveDaoImpl;
import com. entity. Approve;

/**
 * 审批信息服务实现类
 * @author 唐伟
 *
 */
public class ApproveServiceImpl implements ApproveService {
    private ApproveDao cantonDao;

    public ApproveServiceImpl( ) {
        cantonDao = new ApproveDaoImpl( ) ;
    }

    public boolean addApprove( Approve approve)  {
        return false;
    }

    public boolean deleteApprove( String id)  {
        return cantonDao. deleteApprove( id) ;
    }

    public Approve findApprove( int id)  {
        // TODO Auto - generated method stub
        return null;
    }
```

```java
public List < Approve > listApprove( ) {
    return cantonDao.listApprove( );
}

public boolean updateApprove( Approve approve ) {
    // TODO Auto-generated method stub
    return false;
}

}
```

任务九：建设用地审批电子报盘管理系统供地方案模块

一、任务描述

你作为《建设用地审批电子报盘管理系统》项目开发组的程序员，请实现如下功能：
> 供地方案信息的列表显示；
> 供地方案信息的修改。

二、功能描述

（1）点击图9.1中左边导航条中的"供地方案"，则在右边的主体部分显示供地方案信息列表。

图9.1　供地方案信息列表页面

（2）点击图9.1中的"修改"按钮，则进入供地方案信息修改页面，如图9.2所示。
（3）对图9.2中打"＊"号的输入部分进行必填并校验。
（4）点击图9.2中"确定"按钮，在供地方案表中修改一条供地方案信息。
（5）供地方案信息修改成功后，自动定位到供地方案信息列表页面，显示更新后的供地方案信息列表，如图9.1。
（6）测试程序，在供地方案信息页面修改两条以上供地方案信息。

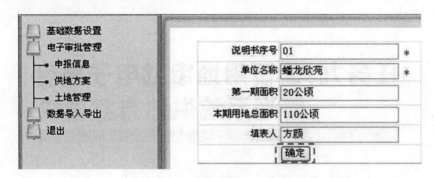

图 9.2　供地方案信息修改页面

三、要求

1. 界面实现

以提供的素材为基础,实现图 9.1、图 9.2 所示页面。

2. 数据库实现

(1)创建数据库 LandDB。

(2)创建供地方案信息表(T_land_offer_scheme),表结构见表 9.1。

表 9.1　供地方案信息表(T_land_offer_scheme)表结构

字段名	字段说明	字段类型	允许为空	备注
Bpl_guid	呈报说明书序号	varchar(38)	否	主键
Unit_name	单位名称	varchar(60)	否	
Period1_area	前期面积	numeric(12,4)	是	数值型,单位:公顷
Per_sum_area	本期用地总面积	numeric(12,4)	是	数值型,单位:公顷
Inputer	填表人	varchar(60)	是	

(3)在表 T_land_offer_scheme 插入记录,见表 9.2。

表 9.2　供地方案信息表(T_land_offer_scheme)记录

Bpl_guid	Unit_name	Period1_area	Per_sum_area	Inputer
0001	蟠龙欣苑	20	40	方颜
0002	藏龙卧虎	50	100	刘芳

3. 功能实现

(1)功能需求如图 9.3 所示。

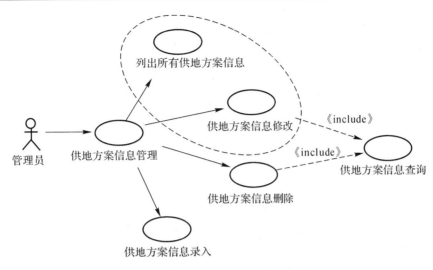

图9.3　供地方案信息设置模块用例图

（2）依据供地方案信息列表活动图完成供地方案信息列表显示功能，如图9.4所示。

（3）依据修改供地方案信息活动图完成修改供地方案功能，如图9.5所示。

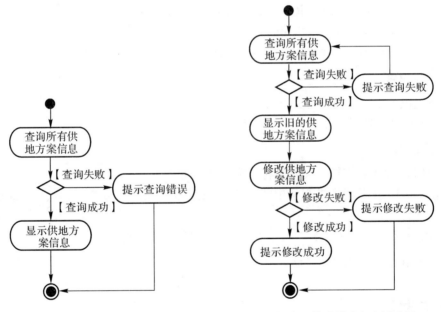

图9.4　供地方案信息列表活动图　　　图9.5　修改供地方案活动图

四、必备知识

1. 数据库相关知识

（1）使用 MS SQL Server 2005/2008 创建数据库，创建数据表，设置表的字段，数据类型，主键，外键，约束。

（2）向数据表插入、删除、修改、查询数据。

2. 页面相关知识

(1) 使用 HTML 制作项目页面。

(2) 使用 CSS 控制页面的样式。

(3) 使用 JavaScript 对页面必要的内容进行校验。

3. JSP 相关知识

(1) 使用 JSTL 标准标签库控制页面显示逻辑。

(2) 理解 JSP 的 request, response, session, application 的概念。

(3) 使用 EL 表达式在页面显示数据。

(4) 合理地使用转发和重定向控制项目的页面跳转。

(5) 使用 JDBC 与数据库进行交互。

(6) MVC 模式下的分层架构，控制器，视图，模型的划分和通信。

五、解题思路

1. 数据库思路

(1) 根据项目要求创建数据库和数据表，向数据表中插入合适的测试数据。

(2) 导入 JDBC 驱动包，编写 JDBC 的连接工具代码。

(3) 编写数据操作对象代码，负责进行与数据库的交互操作。

2. 视图层思路

(1) 将提供的素材页面改写为 JSP 页面。

(2) JSP 使用 JSTL 和 EL 负责控制页面显示的逻辑。

3. 控制层思路

(1) 使用 Servlet 类控制一次请求响应过程的处理。

(2) 由 Servlet 按照顺序进行请求的处理、数据库交互、模型存取和封装、页面跳转逻辑控制等。

4. 模型层思路

使用 JavaBean 作为模型层，封装数据和行为。

六、操作步骤

1. 准备数据库

(1) 根据项目要求，在 SQL Server2008 中创建 LandDB 数据库、供地方案信息表(T_land_offer_scheme)，并插入测试数据。

(2) 编写 JDBC 的连接工具代码。

```
package com.dao;

import java.sql.Connection;
```

```java
import java.sql.DriverManager;
import java.sql.PreparedStatement;
import java.sql.ResultSet;
import java.sql.SQLException;

/**
 * 数据库连接类
 * @author 唐伟
 *
 */
public class DBConnection {
    private static final String DRIVER = "com.microsoft.sqlserver.jdbc.SQLServerDriver";
    private static final String URL = "jdbc:sqlserver://localhost:1433;DatabaseName=LandDB";
    private static final String USER = "sa";
    private static final String PWD = "123";

    private static Connection con;
    private static PreparedStatement prst;
    private static ResultSet rest;

    /**
     * 加载驱动
     */
    static {
        try {
            Class.forName(DRIVER);
        } catch (ClassNotFoundException e) {
            e.printStackTrace();
        }
    }

    /**
     * 取得连接
     * @return
     */
    public static Connection getConnection() {
        try {
            con = DriverManager.getConnection(URL, USER, PWD);
        } catch (SQLException e) {
            e.printStackTrace();
        }
        return con;
    }
```

```java
/**
 * 关闭连接
 */
public static void close() {
    try {
        if (rest != null) {
            rest.close();
        }
        if (prst != null) {
            prst.close();
        }
        if (con != null) {
            con.close();
        }
    } catch (SQLException e) {
        e.printStackTrace();
    }
}
}


package com.dao;

import java.util.List;

import com.entity.Offer;

/**
 * 供地方案操作接口
 * @author 唐伟
 *
 */
public interface OfferDao {
    /**
     * 添加供地方案信息
     * @param canton
     * @return
     */
    boolean addOffer(Offer offers);

    /**
     * 修改供地方案信息
     * @param canton
```

```java
     * @return
     */
    boolean updateOffer(Offer offer);

    /**
     * 删除供地方案信息
     * @param id
     * @return
     */
    boolean deleteOffer(int id);

    /**
     * 查询所有的供地方案信息
     * @return
     */
    List<Offer> listOffer();

    /**
     * 查询供地方案信息
     * @param id
     * @return
     */
    Offer findOffer(String id);
}


package com.dao;

import java.sql.Connection;
import java.sql.PreparedStatement;
import java.sql.ResultSet;
import java.sql.SQLException;
import java.util.ArrayList;
import java.util.List;

import com.entity.Offer;

/**
 * 供地方案操作实现类
 * @author 唐伟
 *
 */
public class OfferDaoImpl implements OfferDao {
```

```java
public boolean addOffer(Offer offer) {
    return false;
}

public boolean deleteOffer(int id) {
    // TODO Auto-generated method stub
    return false;
}

public Offer findOffer(String id) {
    Offer offer = new Offer();
    String sql = "select * from T_LAND_OFFER_SCHEME where BPL_GUID = ?";
    Connection con = null;
    PreparedStatement prst = null;
    ResultSet rest = null;
    con = DBConnection.getConnection();
    try {
        prst = con.prepareStatement(sql);
        prst.setString(1, id);
        rest = prst.executeQuery();
        if(rest.next()) {
            offer.setBpl_guid(rest.getString("bpl_guid"));
            offer.setUnti_name(rest.getString("unit_name"));
            offer.setPeriod1_area(rest.getDouble("period1_Area"));
            offer.setPer_sum_area(rest.getDouble("per_Sum_Area"));
            offer.setInputer(rest.getString("inputer"));
        }
    } catch (SQLException e) {
        e.printStackTrace();
    }
    return offer;
}

public List<Offer> listOffer() {
    List<Offer> list = new ArrayList<Offer>();
    String sql = "select * from T_LAND_OFFER_SCHEME";
    Connection con = null;
    PreparedStatement prst = null;
    ResultSet rest = null;
    con = DBConnection.getConnection();
    try {
        prst = con.prepareStatement(sql);
        rest = prst.executeQuery();
```

```java
            while(rest.next()){
                Offer offer = new Offer();
                offer.setBpl_guid(rest.getString("bpl_guid"));
                offer.setUnti_name(rest.getString("unit_name"));
                offer.setPeriod1_area(rest.getDouble("period1_Area"));
                offer.setPer_sum_area(rest.getDouble("per_Sum_Area"));
                offer.setInputer(rest.getString("inputer"));
                list.add(offer);
            }
        } catch (SQLException e) {
            e.printStackTrace();
        }
        return list;
    }


    public boolean updateOffer(Offer offer) {
        boolean flag = false;
        String sql = "update T_LAND_OFFER_SCHEME set UNIT_NAME = ?, Period1_Area = ?, Per_Sum_Area = ?, inputer = ? where BPL_GUID = ?";
        Connection con = null;
        PreparedStatement prst = null;
        con = DBConnection.getConnection();
        try {
            prst = con.prepareStatement(sql);
            prst.setString(1, offer.getUnti_name());
            prst.setDouble(2, offer.getPeriod1_area());
            prst.setDouble(3, offer.getPer_sum_area());
            prst.setString(4, offer.getInputer());
            prst.setString(5, offer.getBpl_guid());
            if(prst.executeUpdate() != 0) {
                flag = true;
            }
        } catch (SQLException e) {
            e.printStackTrace();
        } finally {
            DBConnection.close();
        }
        return flag;
    }

}
```

2. 编写视图层代码

```jsp
<%@ page language="java" import="java.util.*" pageEncoding="gb2312"%>
```

```jsp
<%@ taglib uri="http://java.sun.com/jsp/jstl/core" prefix="c" %>
<%
String path = request.getContextPath();
String basePath = request.getScheme() + "://" + request.getServerName() + ":" + request.getServerPort() + path + "/";
%>

<!DOCTYPE HTML PUBLIC "-//W3C//DTD HTML 4.01 Transitional//EN">
<html>
  <head>
    <base href="<%=basePath%>">

    <title>My JSP 'listOffer.jsp' starting page</title>

    <meta http-equiv="pragma" content="no-cache">
    <meta http-equiv="cache-control" content="no-cache">
    <meta http-equiv="expires" content="0">
    <meta http-equiv="keywords" content="keyword1,keyword2,keyword3">
    <meta http-equiv="description" content="This is my page">
    <!--
    <link rel="stylesheet" type="text/css" href="styles.css">
    -->

  </head>

  <body>
    <c:choose>
    <c:when test="${empty list}">
       <center><p>没有供地方案信息</p></center>
    </c:when>
    <c:otherwise>
    <table width="495" border="1" style="border-collapse:collapse">
     <tr>
       <td width="72"><div align="center"><font color="#3074A2" style="font-size:9pt;color:#000000">说明书序号</font></div></td>
       <td width="72"><div align="center"><font color="#3074A2" style="font-size:9pt;color:#000000">单位名称</font></div></td>
       <td width="72"><div align="center"><font color="#3074A2" style="font-size:9pt;color:#000000">第一期面积</font></div></td>
       <td width="95"><div align="center"><font color="#3074A2" style="font-size:9pt;color:#000000">本期用地总面积</font></div></td>
       <td width="46"><div align="center"><font color="#3074A2" style="font-size:9pt;color:#000000">填表人</font></div></td>
```

```
                <td width="105"><div align="center"></div></td>
            </tr>
            <c:forEach var="offer" items="${list}">
            <tr>
                <td><div align="center"><font color="#3074A2" style="font-size:9pt;color:#000000">${offer.bpl_guid}</font></div></td>
                <td><div align="center"><font color="#3074A2" style="font-size:9pt;color:#000000">${offer.unti_name}</font></div></td>
                <td><div align="center"><font color="#3074A2" style="font-size:9pt;color:#000000">${offer.period1_area}</font></div></td>
                <td><div align="center"><font color="#3074A2" style="font-size:9pt;color:#000000">${offer.per_sum_area}</font></div></td>
                <td><div align="center"><font color="#3074A2" style="font-size:9pt;color:#000000">${offer.inputer}</font></div></td>
                <td>
                <input type="button" name="bt1" value="删除">
                <a href="updateOffer.html" target="_blank"><input type="button" name="bt2" value="修改" onClick="javascript:window.location.href='OfferSearchServlet?bpl_guid=${offer.bpl_guid}'"></a>
                </td>
            </tr>
            </c:forEach>
        </table>
        </c:otherwise>
    </c:choose>
    </body>
</html>


<%@ page language="java" import="java.util.*" pageEncoding="gb2312"%>
<%
String path = request.getContextPath();
String basePath = request.getScheme()+"://"+request.getServerName()+":"+request.getServerPort()+path+"/";
%>

<!DOCTYPE HTML PUBLIC "-//W3C//DTD HTML 4.01 Transitional//EN">
<html>
    <head>
        <base href="<%=basePath%>">

        <title>My JSP 'updateOffer.jsp' starting page</title>

        <meta http-equiv="pragma" content="no-cache">
```

```html
<meta http-equiv="cache-control" content="no-cache">
<meta http-equiv="expires" content="0">
<meta http-equiv="keywords" content="keyword1,keyword2,keyword3">
<meta http-equiv="description" content="This is my page">
<!--
<link rel="stylesheet" type="text/css" href="styles.css">
-->

</head>
<script language="javascript">
function check(){
    if(form1.bpl_guid.value==""){
        alert("请输入说明书序号");
        form1.bpl_guid.focus();
        return false;
    }
    if(form1.unti_name.value==""){
        alert("请输入单位名称");
        form1.unti_name.focus();
        return false;
    }
}
</script>
<body>
    <table width="340" border="0">
<tr><td width="553"><form action="servlet/OfferUpdateServlet" method="post" name="form1" onSubmit="return check()">
    <table width="290" height="70" border="1" align="center" style="border-collapse:collapse">
        <tr>
            <td width="131" height="9"><div align="right"><font color="#3074A2" style="font-size:9pt;color:#000000">说明书序号</font></div></td>
            <td width="188"><input name="bpl_guid" type="text" id="bpl_guid" size="20" value="${offer.bpl_guid}" readonly="true"></td>
        </tr>
        <tr>
            <td width="131" height="4"><div align="right"><font color="#3074A2" style="font-size:9pt;color:#000000">单位名称</font></div></td>
            <td><input name="unti_name" type="text" id="unti_name" size="20" value="${offer.unti_name}"></td>
        </tr>
        <tr>
```

```html
<td width="131" height="29"><div align="right"><font color="#3074A2" style="font-size:9pt;color:#000000">第一期面积</font></div></td>
<td><input name="period1_area" type="text" id="period1_area" size="20" value="${offer.period1_area}"></td>
</tr>
<tr>
<td width="131" height="4"><div align="right"><font color="#3074A2" style="font-size:9pt;color:#000000">本期用地总面积</font></div></td>
<td><input name="per_sum_area" type="text" id="per_sum_area" size="20" value="${offer.per_sum_area}"></td>
</tr>
<tr>
<td height="4"><div align="right"><font color="#3074A2" style="font-size:9pt;color:#000000">填表人</font></div></td>
<td><input name="inputer" type="text" id="inputer" size="20" value="${offer.inputer}"></td>
</tr>
<tr>
<td width="131" height="4"><div align="right"></div></td>
<td><input type="submit" name="bt1" value="确定"></td>
</tr>
</table>
</form></td>
</tr>
</table>
</body>
</html>
```

3. 编写模型层代码

```java
package com.entity;

/**
 * 供地方案信息
 * @author 唐伟
 *
 */
public class Offer {
    private String bpl_guid;//呈报说明书序号
    private String unti_name;//单位名称
    private double period1_area;//第一期面积
    private double per_sum_area;//本期用地总面积
    private String inputer;//填表人
```

```java
public String getBpl_guid() {
    return bpl_guid;
}
public void setBpl_guid(String bpl_guid) {
    this.bpl_guid = bpl_guid;
}
public String getUnti_name() {
    return unti_name;
}
public void setUnti_name(String unti_name) {
    this.unti_name = unti_name;
}
public double getPeriod1_area() {
    return period1_area;
}
public void setPeriod1_area(double period1_area) {
    this.period1_area = period1_area;
}
public double getPer_sum_area() {
    return per_sum_area;
}
public void setPer_sum_area(double per_sum_area) {
    this.per_sum_area = per_sum_area;
}
public String getInputer() {
    return inputer;
}
public void setInputer(String inputer) {
    this.inputer = inputer;
}
```

}

4. 编写控制层代码

```java
package com.servlet;

import java.io.IOException;
import java.io.PrintWriter;
import java.util.List;

import javax.servlet.ServletException;
import javax.servlet.http.HttpServlet;
import javax.servlet.http.HttpServletRequest;
```

```java
import javax.servlet.http.HttpServletResponse;

import com.entity.Offer;
import com.service.OfferService;
import com.service.OfferServiceImpl;

/**
 * 供地方案查询所有控制类
 * @author 唐伟
 *
 */
public class OfferListServlet extends HttpServlet {
    private OfferService offerService;
    /**
     * The doGet method of the servlet. <br>
     *
     * This method is called when a form has its tag value method equals to get.
     *
     * @param request the request send by the client to the server
     * @param response the response send by the server to the client
     * @throws ServletException if an error occurred
     * @throws IOException if an error occurred
     */
    public void doGet(HttpServletRequest request, HttpServletResponse response)
            throws ServletException, IOException {

        this.doPost(request, response);
    }

    /**
     * The doPost method of the servlet. <br>
     *
     * This method is called when a form has its tag value method equals to post.
     *
     * @param request the request send by the client to the server
     * @param response the response send by the server to the client
     * @throws ServletException if an error occurred
     * @throws IOException if an error occurred
     */
    public void doPost(HttpServletRequest request, HttpServletResponse response)
            throws ServletException, IOException {

        response.setContentType("text/html");
```

```java
            PrintWriter out = response.getWriter();
            offerService = new OfferServiceImpl();
            List<Offer> list = offerService.listOffer();
            request.setAttribute("list", list);
            request.getRequestDispatcher("../listOffer.jsp").forward(request, response);
            out.flush();
            out.close();
    }
}


package com.servlet;

import java.io.IOException;
import java.io.PrintWriter;

import javax.servlet.ServletException;
import javax.servlet.http.HttpServlet;
import javax.servlet.http.HttpServletRequest;
import javax.servlet.http.HttpServletResponse;

import com.entity.Offer;
import com.service.OfferService;
import com.service.OfferServiceImpl;

/**
 * 供地方案查询控制类
 * @author 唐伟
 *
 */
public class OfferSearchServlet extends HttpServlet {
    private OfferService offerService;
    /**
     * The doGet method of the servlet. <br>
     *
     * This method is called when a form has its tag value method equals to get.
     *
     * @param request the request send by the client to the server
     * @param response the response send by the server to the client
     * @throws ServletException if an error occurred
     * @throws IOException if an error occurred
     */
    public void doGet(HttpServletRequest request, HttpServletResponse response)
```

```java
        throws ServletException, IOException {
    this.doPost(request, response);
}

/**
 * The doPost method of the servlet. <br>
 *
 * This method is called when a form has its tag value method equals to post.
 *
 * @param request the request send by the client to the server
 * @param response the response send by the server to the client
 * @throws ServletException if an error occurred
 * @throws IOException if an error occurred
 */
public void doPost(HttpServletRequest request, HttpServletResponse response)
        throws ServletException, IOException {

    response.setContentType("text/html");
    PrintWriter out = response.getWriter();
    String id = request.getParameter("bpl_guid");
    offerService = new OfferServiceImpl();
    Offer offer = offerService.findOffer(id);
    request.setAttribute("offer", offer);
    request.getRequestDispatcher("../updateOffer.jsp").forward(request, response);
    out.flush();
    out.close();
}
}



package com.servlet;

import java.io.IOException;
import java.io.PrintWriter;

import javax.servlet.ServletException;
import javax.servlet.http.HttpServlet;
import javax.servlet.http.HttpServletRequest;
import javax.servlet.http.HttpServletResponse;

import com.entity.Offer;
import com.service.OfferService;
```

```java
import com.service.OfferServiceImpl;

/**
 * 供地方案添加控制类
 * @author 唐伟
 *
 */
public class OfferUpdateServlet extends HttpServlet {
    private OfferService offerService;
    /**
     * The doGet method of the servlet. <br>
     *
     * This method is called when a form has its tag value method equals to get.
     *
     * @param request the request send by the client to the server
     * @param response the response send by the server to the client
     * @throws ServletException if an error occurred
     * @throws IOException if an error occurred
     */
    public void doGet(HttpServletRequest request, HttpServletResponse response)
            throws ServletException, IOException {

        this.doPost(request, response);
    }

    /**
     * The doPost method of the servlet. <br>
     *
     * This method is called when a form has its tag value method equals to post.
     *
     * @param request the request send by the client to the server
     * @param response the response send by the server to the client
     * @throws ServletException if an error occurred
     * @throws IOException if an error occurred
     */
    public void doPost(HttpServletRequest request, HttpServletResponse response)
            throws ServletException, IOException {

        response.setContentType("text/html");
        PrintWriter out = response.getWriter();
        String bpl_guid = request.getParameter("bpl_guid");
        String unti_name = request.getParameter("unti_name");
        double period1_area = Double.parseDouble(request.getParameter("period1_area"));
```

```java
            double per_sum_area = Double.parseDouble(request.getParameter("per_sum_area"));
            String inputer = request.getParameter("inputer");
            Offer offer = new Offer();
            offer.setBpl_guid(bpl_guid);
            offer.setUnti_name(unti_name);
            offer.setPeriod1_area(period1_area);
            offer.setPer_sum_area(per_sum_area);
            offer.setInputer(inputer);
            offerService = new OfferServiceImpl();
            if(offerService.updateOffer(offer)) {
                request.getRequestDispatcher("OfferListServlet").forward(request, response);
            } else {
                System.out.println("fail");
                request.getRequestDispatcher("../error.html").forward(request, response);
            }
            out.flush();
            out.close();
        }
}


package com.service;

import java.util.List;

import com.entity.Offer;

/**
 * 供地方案服务接口
 * @author 唐伟
 *
 */
public interface OfferService {
    /**
     * 添加供地方案信息
     * @param canton
     * @return
     */
    boolean addOffer(Offer offer);

    /**
     * 修改供地方案信息
     * @param canton
```

```java
     * @return
     */
    boolean updateOffer(Offer offer);

    /**
     * 删除供地方案信息
     * @param id
     * @return
     */
    boolean deleteOffer(int id);

    /**
     * 查询所有的供地方案信息
     * @return
     */
    List<Offer> listOffer();

    /**
     * 查询供地方案信息
     * @param id
     * @return
     */
    Offer findOffer(String id);
}
```

```java
package com.service;

import java.util.List;

import com.dao.OfferDao;
import com.dao.OfferDaoImpl;
import com.entity.Offer;

/**
 * 供地方案服务实现类
 * @author 唐伟
 *
 */
public class OfferServiceImpl implements OfferService {
    private OfferDao offerDao;

    public OfferServiceImpl() {
        offerDao = new OfferDaoImpl();
```

```java
    }

    public boolean addOffer(Offer offer) {
        return false;
    }

    public boolean deleteOffer(int id) {
        // TODO Auto-generated method stub
        return false;
    }

    public Offer findOffer(String id) {
        return offerDao.findOffer(id);
    }

    public List<Offer> listOffer() {
        return offerDao.listOffer();
    }

    public boolean updateOffer(Offer offer) {
        return offerDao.updateOffer(offer);
    }

}
```

任务十：网上书店图书信息模块

一、任务描述

你作为《网上书店》项目开发组的程序员，请实现如下功能：
- 查看所有图书信息；
- 添加图书信息。

二、功能描述

（1）点击图 10.1 所示页面左边导航条中的"查看图书"菜单项，则在右边的主体部分显示图书信息列表。

图 10.1　网上书店页面

(2)点击图10.1所示页面左边导航条中的"增加图书"菜单项,则进入图书信息录入页面,如图10.2所示。

图10.2 网上书店书籍信息录入页面

(3)点击图10.2中的"添加"按钮,对图中打"＊"号的输入部分进行必填校验,通过校验后在数据库中添加图书信息。

(4)图书信息添加成功后,跳转到图10.1所示页面,显示更新后的图书信息列表。

(5)测试程序,在添加图书信息页面新增两条以上图书信息。

三、要求

1. 界面实现

以提供的素材为基础,实现图10.1、图10.2所示页面。

2. 数据库实现

(1)创建数据库 BookStoreDB。

(2)创建图书信息表(T_book),表结构见表10.1。

表10.1 图书信息表(T_book)表结构

字段名	字段说明	字段类型	允许为空	备注
Book_number	图书编号	varchar(45)	否	主键
Book_name	图书名称	varchar(64)	否	
Author	作者	varchar(30)	否	
Publisher	出版社	varchar(30)	否	
Price	价格	float	否	数值型

(3) 在表 T_book 插入以下记录,见表 10.2。

表 10.2　图书信息表(T_book)记录

Book_number	Book_name	Author	Publisher	Price
000001	JSP	王红	北京大学出版社	44
000002	Linux	刘威	西安电子出版社	39

3. 功能实现

(1) 功能需求如图 10.3 所示。

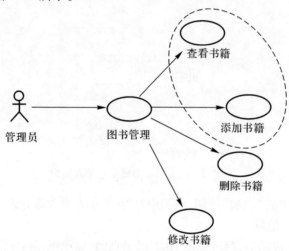

图 10.3　图书管理模块用例图

(2) 依据查看图书信息活动图完成图书信息查看显示功能,如图 10.4 所示。

(3) 依据增加图书信息活动图完成增加图书信息功能,如图 10.5 所示。

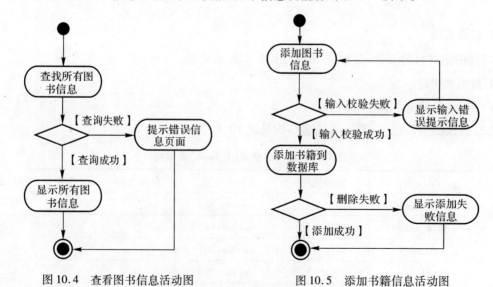

图 10.4　查看图书信息活动图　　　　图 10.5　添加书籍信息活动图

四、必备知识

1. 数据库相关知识

(1) 使用 MS SQL Server 2005/2008 创建数据库,创建数据表,设置表的字段,数据类型,主键,外键,约束。

(2) 向数据表插入、删除、修改、查询数据。

2. 页面相关知识

(1) 使用 HTML 制作项目页面。

(2) 使用 CSS 控制页面的样式。

(3) 使用 JavaScript 对页面必要的内容进行校验。

3. JSP 相关知识

(1) 使用 JSTL 标准标签库控制页面显示逻辑。

(2) 理解 JSP 的 request,response,session,application 的概念。

(3) 使用 EL 表达式在页面显示数据。

(4) 合理地使用转发和重定向控制项目的页面跳转。

(5) 使用 JDBC 与数据库进行交互。

(6) MVC 模式下的分层架构,控制器,视图,模型的划分和通信。

五、解题思路

1. 数据库思路

(1) 根据项目要求创建数据库和数据表,向数据表中插入合适的测试数据。

(2) 导入 JDBC 驱动包,编写 JDBC 的连接工具代码。

(3) 编写数据操作对象代码,负责进行与数据库的交互操作。

2. 视图层思路

(1) 将提供的素材页面改写为 JSP 页面。

(2) JSP 使用 JSTL 和 EL 负责控制页面显示的逻辑。

3. 控制层思路

(1) 使用 Servlet 类控制一次请求响应过程的处理。

(2) 由 Servlet 按照顺序进行请求的处理、数据库交互、模型存取和封装、页面跳转逻辑控制等。

4. 模型层思路

使用 JavaBean 作为模型层,封装数据和行为。

六、操作步骤

1. 准备数据库

（1）根据项目要求，在 SQL Server2008 中创建 BookStoreDB 数据库、图书信息表（T_book），并插入测试数据。

（2）编写 JDBC 的连接工具代码。

```java
package bean;

import java.sql.*;

public class DBUtil
{
    private static String url = "jdbc:mysql://localhost:3306/bookstoredb?useUnicode=true&characterEncoding=gbk";
    private static String user = "root";
    private static String password = "root";

    /**
     * 得到一个 Connection 对象
     * @return java.sql.Connection
     */
    public static Connection getConnection()
    {
        Connection conn = null;
        try
        {
            Class.forName("com.mysql.jdbc.Driver");
            conn = DriverManager.getConnection(url, user, password);
        }
        catch(ClassNotFoundException e)
        {
            e.printStackTrace();
        }
        catch(SQLException e1)
        {
            e1.printStackTrace();
        }
        return conn;
    }
    /**
     * 关闭指定的结果集
```

```java
         * @param rs 要关闭的结果集
         */
        public static void closeResultSet(ResultSet rs)
        {
            if(rs != null)
            {
                try
                {
                    rs.close();
                }
                catch(SQLException e)
                {
                }
            }
        }
        /**
         * 关闭指定的 Statement
         * @param stmt 要关闭的 Statement
         */
        public static void closeStatement(Statement stmt)
        {
            if(stmt != null)
            {
                try
                {
                    stmt.close();
                }
                catch(SQLException e)
                {
                }
            }
        }
        /**
         * 关闭连接
         * @param conn 要关闭的连接
         */
        public static void closeConnection(Connection conn)
        {
            if(conn != null)
            {
                try
                {
                    conn.close();
```

```
                    }
                    catch(SQLException e)
                    {
                    }
                }
        }
        public static void main(String[] args)
        {
            System.out.println(getConnection());
        }
}


package bean;

import java.io.Serializable;
import java.util.ArrayList;
import java.util.Collection;
import java.util.Iterator;
import java.sql.*;

import javax.sql.DataSource;
import javax.naming.Context;
import javax.naming.InitialContext;
import javax.naming.NamingException;

public class BookDBBean implements Serializable
{
    private static String url = "jdbc:mysql://localhost:3306/bookstoredb";
    private static String user = "root";
    private static String password = "root";

    /**
     * 得到一个 Connection 对象
     * @return java.sql.Connection
     */
    public static Connection getConnection()
    {
        Connection conn = null;
        try
        {
            Class.forName("com.mysql.jdbc.Driver");
            conn = DriverManager.getConnection(url, user, password);
        }
```

```java
        catch(ClassNotFoundException e)
        {
            e.printStackTrace();
        }
        catch(SQLException e1)
        {
            e1.printStackTrace();
        }
        return conn;
    }
    /**
     * 关闭连接对象。
     */
    protected void closeConnection(Connection conn)
    {
        if(conn! = null)
        {
            try
            {
                conn.close();
                conn = null;
            }
            catch (SQLException ex)
            {
                ex.printStackTrace();
            }
        }
    }

    /**
     * 关闭Statement对象。
     */
    protected void closeStatement(Statement stmt)
    {
        if(stmt! = null)
        {
            try
            {
                stmt.close();
                stmt = null;
            }
            catch (SQLException ex)
            {
```

```java
                ex.printStackTrace();
            }
        }
    }

    /**
     * 关闭 PreparedStatement 对象。
     */
    protected void closePreparedStatement(PreparedStatement pstmt)
    {
        if(pstmt! = null)
        {
            try
            {
                pstmt.close();
                pstmt = null;
            }
            catch (SQLException ex)
            {
                ex.printStackTrace();
            }
        }
    }

    /**
     * 关闭 ResultSet 对象。
     */
    protected void closeResultSet(ResultSet rs)
    {
        if(rs! = null)
        {
            try
            {
                rs.close();
                rs = null;
            }
            catch (SQLException ex)
            {
                ex.printStackTrace();
            }
        }
    }
```

```java
/**
 *得到数据库中所有的图书信息。
 */
public Collection getBooks( ) throws SQLException
{
    Connection conn = null;
    Statement stmt = null;
    ResultSet rs = null;
    ArrayList   bookList = new ArrayList( );

    try
    {
        conn = getConnection( );
        stmt = conn.createStatement( );
        rs = stmt.executeQuery("select * from t_book");
        while (rs.next( ))
        {
            BookBean book = new BookBean(rs.getString(1), rs.getString(2), rs.getString(3), rs.getString(4), rs.getFloat(6));
            bookList.add(book);
        }
        return bookList;
    }
    finally
    {
        closeResultSet(rs);
        closeStatement(stmt);
        closeConnection(conn);
    }
}

/**
 *得到选择的图书信息。
 */
public BookBean getBook(String Book_number) throws SQLException
{
    Connection conn = null;
    PreparedStatement pstmt = null;
    ResultSet rs = null;

    try
    {
```

```
            conn = getConnection( );
            pstmt = conn.prepareStatement("select * from t_book where Book_number = ?");
            pstmt.setString(1, Book_number);
            rs = pstmt.executeQuery( );
            BookBean book = null;
            if(rs.next( ))
            {
                book = new BookBean(rs.getString(1), rs.getString(2), rs.getString(3),
                        rs.getString(4), rs.getFloat(6));
            }
            return book;
        }
        finally
        {
            closeResultSet(rs);
            closePreparedStatement(pstmt);
            closeConnection(conn);
        }
    }
}
```

2. 编写视图层代码

```
<%@ page language="java" pageEncoding="gbk"%>
<%@ taglib prefix="c" uri="http://java.sun.com/jsp/jstl/core" prefix="c"%>
<!DOCTYPE html PUBLIC "-//W3C//DTD XHTML 1.0 Transitional//EN" "http://www.w3.org/TR/xhtml1/DTD/xhtml1-transitional.dtd">
    <script language"javascript">
        function init( ){
            alert("${info}");
        }
        <c:if test="${!empty info}">
        window.onload = init;
        </c:if>
    </script>
    <html>
    <head>
    <title>413 网上书城|管理员</title>
    <link href="22.css" rel="stylesheet" type="text/css"/>
    </head>

    <body>
    <div id="container">
```

```html
<div id="banner"><img src="pic1/p01.jpg"/></div>
<div id="globallink">
    <ul>
        <li><a href="enter.jsp">首页</a></li>
        <li><a href="login.jsp">用户注册</a></li>
        <li><a href="hot_books.jsp">图书促销</a></li>
        <li><a href="#">在线阅读</a></li>
        <li><a href="vip.jsp">会员中心</a></li>
        <li><a href="about_me.jsp">联系我们</a></li>
        <li><a href="#">支付方式</a></li>
    </ul>
    <br/>
</div>
<div id="left">
  <div id="login">
    网上书店欢迎你  <a href="booklist.jsp">[退出]</a>
    <ul>
    <li>图书信息管理
      <ul>
        <li><a href="FindAllBook">查看图书</a></li>
        <li><a href="addbook.jsp">增加图书</a></li>
      </ul>
    </li>
    <li><a href="#">订单管理</a></li>
    <li><a href="#">客户管理</a></li>
    <li><a href="#">查看留言</a></li>
    </ul>
  </div>
  <div id="category">
    <h4><span>种类</span></h4>
    <ul>
        <li><a href="#">计算机/网络</a></li>
        <li><a href="#">人文科社</a></li>
        <li><a href="#">培训教材</a></li>
        <li><a href="#">经济</a></li>
        <li><a href="#">管理</a></li>
        <li><a href="#">文化/文学</a></li>
        <li><a href="#">百科全书</a></li>
        <li><a href="#">艺术</a></li>
        <li><a href="#">小说</a></li>
        <li><a href="#">期刊杂志</a></li>
    </ul>
    <h4><span>品牌出版社</span></h4>
```

```
<ul>
    <li><a href="#">人民教育出版社</a></li>
    <li><a href="#">现代出版社</a></li>
    <li><a href="#">新世纪出版社</a></li>
    <li><a href="#">中国少儿出版社</a></li>
    <li><a href="#">二十一世纪出版社</a></li>
    <li><a href="#">清华大学出版社</a></li>
    <li><a href="#">泰博文化出版社</a></li>
    <li><a href="#">科学普及出版社</a></li>
</ul>

        </div>
    </div>
    <div id="main">
        <div id="latest">
<br/><br/><br/><br/>
<h3 align="center"><font color="#ff0000">图书信息列表</font></h3>
<font size=2>
    共有${pageCount}页,这是第${pageNo}页。<!--如果是第一页,则不显示超链接-->
        <c:if test="${pageNo==1}">
            第一页
            上一页
        </c:if> <!--如果不是第一页,则显示超链接-->
    <c:if test="${pageNo!=1}">
        <a href="FindAllBook?pageNo=1">第一页</a>
        <a href="FindAllBook?pageNo=${pageNo-1}">上一页</a>
    </c:if> <!--如果是最后一页,则不显示超链接-->
    <c:if test="${pageNo==pageCount}">
        下一页
        最后一页
    </c:if> <!--如果不是最后一页,则显示超链接-->
<c:if test="${pageNo!=pageCount}">
        <a href="FindAllBook?pageNo=${pageNo+1}">下一页</a>
        <a href="FindAllBook?pageNo=${pageCount}">最后一页</a>
    </c:if>
<form action="FindAllBook">
跳转到<input type="text" name="pageNo">页<input type="submit" value="跳转">
</form>
</font>
<table align="center" border="1">
    <tr>
        <th align="center" font="bold">
            图书编号
```

```html
            </th>
            <th align="center" font="bold">
                图书标题
            </th>
            <th align="center" font="bold">
                作者
            </th>
            <th align="center" font="bold">
                出版社
            </th>
            <th align="center" font="bold">
                价格
            </th>
            <th align="center" colspan="2" font="bold">
                功能
            </th>
        </tr>
        <c:forEach items="${booklist}" var="book">
            <tr>
                <td>
                    ${book.book_number}
                </td>
                <td>
                    ${book.book_name}
                </td>

                <td>
                    ${book.author}
                </td>
                <td>
                    ${book.publisher}
                </td>

                <td>
                    ${book.price}
                </td>

                <td>
<form action="UpdateFindBook" method="post">
<input type="submit" value="修改"/>
<input type="hidden" name="Book_number" value="${book.book_number}"/>
```

```
        </form>
      </td>
      <td>
        <form action="DeleteBook" method="post" onSubmit="return confirm('确认要删除该图书信息吗?');">
          <input type="hidden" name="Book_number" value="${book.book_number}"/>
          <input type="submit" value="删除"/>
        </form>
      </td>
    </tr>
  </c:forEach>
</table>
</html>

<%@ page language="java" import="java.util.*" pageEncoding="gbk"%>
<%@ taglib prefix="c" uri="http://java.sun.com/jsp/jstl/core" prefix="c"%>
<!DOCTYPE html PUBLIC "-//W3C//DTD XHTML 1.0 Transitional//EN" "http://www.w3.org/TR/xhtml1/DTD/xhtml1-transitional.dtd">
<script language="javascript">
function isValidate(form1)
{
    Book_name = form1.Book_name.value;
    Book_number = form1.Book_number.value;
    Author = form1.Author.value;
    Publisher = form1.Publisher.value;
    Price = form1.Price.value;

    if(Book_name=="")
    {
        alert("图书名称不能为空!");
        form1.Book_name.focus();
        return false;
    }
    if(Book_number=="")
    {
        alert("图书编号不能为空!");
        form1.Book_number.focus();
        return false;
    }
    if(Author=="")
    {
        alert("作者信息不能为空!");
        form1.Author.focus();
```

```
            return false;
        }
        if(Publisher=="")
        {
            alert("出版社信息不能为空!");
            form1.Publisher.focus();
            return false;
        }
    if(Price=="")
        {
            alert("单价信息不能为空!");
            form1.Price.focus();
    return false;
    }
function isNull(str)
{
        if(str.length==0)
        return true;
        else
        return false;
        }

}
</script>
<html>
<head>
<title>413 网上书城|管理员|添加新图书</title>
<link href="25.css" rel="stylesheet" type="text/css"/>
</head>

<body>
    <div id="container">
    <div id="banner"><img src="pic1/p01.jpg"/></div>
    <div id="globallink">
        <ul>
            <li><a href="enter.jsp">首页</a></li>
            <li><a href="login.jsp">用户注册</a></li>
            <li><a href="hot_books.jsp">图书促销</a></li>
            <li><a href="#">在线阅读</a></li>
            <li><a href="vip.jsp">会员中心</a></li>
            <li><a href="about_me.jsp">联系我们</a></li>
            <li><a href="#">支付方式</a></li>
```

```html
        </ul>
        <br/>
</div>
<div id="left">
    <div id="login">
        网上书店欢迎你<a href="booklist.jsp">[退出]</a>
        <ul>
            <li>图书信息管理
                <ul>
                    <li><a href="FindAllBook">查看图书</a></li>
                    <li><a href="addbook.jsp">增加图书</a></li>
                </ul>
            </li>
            <li><a href="#">订单管理</a></li>
            <li><a href="#">客户管理</a></li>
            <li><a href="#">查看留言</a></li>
        </ul>
    </div>
    <div id="category">
        <h4><span>种类</span></h4>
        <ul>
            <li><a href="#">计算机/网络</a></li>
            <li><a href="#">人文科社</a></li>
            <li><a href="#">培训教材</a></li>
            <li><a href="#">经济</a></li>
            <li><a href="#">管理</a></li>
            <li><a href="#">文化/文学</a></li>
            <li><a href="#">百科全书</a></li>
            <li><a href="#">艺术</a></li>
            <li><a href="#">小说</a></li>
            <li><a href="#">期刊杂志</a></li>
        </ul>
        <h4><span>品牌出版社</span></h4>
        <ul>
            <li><a href="#">人民教育出版社</a></li>
            <li><a href="#">现代出版社</a></li>
            <li><a href="#">新世纪出版社</a></li>
            <li><a href="#">中国少儿出版社</a></li>
            <li><a href="#">清华大学出版社</a></li>
            <li><a href="#">泰博文化出版社</a></li>
            <li><a href="#">科学普及出版社</a></li>
        </ul>
```

```html
        </div>
      </div>
      <div id="main">
        <div id="latest"><a href="#"><img src="pic1/p06.jpg"/></a></div>
          <form name="form1" action="AddBook" method="post" onsubmit="return isValidate(form1)">
            <div id="latt">
              <h2 align="center">添加图书信息</h2>
  <table align="center">
    <tr>
      <th><font size="4">书名:</font></th>
      <td><input type="text" name="Book_name"/><font size="5" color="red">*</font></td>
    </tr>
    <tr>
      <th><font size="4">书号:</font></th>
      <td><input type="text" name="Book_number"/><font size="5" color="red">*</font></td>
    </tr>
    <tr>
      <th><font size="4">作者:</font></th>
      <td><input type="text" name="Author"/><font size="5" color="red">*</font></td>
    </tr>
    <tr>
      <th><font size="4">出版社:</font></th>
      <td><input type="text" name="Publisher"/><font size="5" color="red">*</font></td>
    </tr>
    <tr>
      <th><font size="4">定价:</font></th>
      <td><input type="text" name="Price"/><font size="5" color="red">*</font></td>
    </tr>

    <tr>
      <td><h4 align="center"><input type="reset" value="重置"/></h4></td>
      <td><h4 align="center"><input type="submit" value="添加"/></h4></td>
    </tr>
  </table>
            </div>
          </form>
          <div id="latest"><a href="#"><img src="pic1/p105.jpg" width="518" height="
```

```
228"/></a></div>
      </div>
        <div id="bottom">
          <p> <a href="E—mail:lvzb212@demo.com">lvzb212@demo.com</a> </p>
        </div>
    </div>
  </body>
</html>

<%@ page language="java" pageEncoding="gbk"%>
<%@ taglib prefix="c" uri="http://java.sun.com/jsp/jstl/core" prefix="c"%>
<% String path = request.getContextPath();
String basePath = request.getScheme()+"://"+request.getServerName()+":"+request.getServerPort()+path+"/";%>
<!DOCTYPE html PUBLIC "-//W3C//DTD XHTML 1.0 Transitional//EN" "http://www.w3.org/TR/xhtml1/DTD/xhtml1-transitional.dtd">
<script language="javascript">
function isValidate(form1)
{
  Book_name=form1.Book_name.value;
  Book_number=form1.Book_number.value;
  Author=form1.Author.value;
  Publisher=form1.Publisher.value;
  Price=form1.Price.value;

  if(Book_name=="")
  {
    alert("图书名称不能为空!");
    form1.Book_name.focus();
    return false;
  }
  if(Book_number=="")
  {
    alert("图书编号不能为空!");
    form1.Book_number.focus();
    return false;
  }
  if(Author=="")
  {
    alert("作者信息不能为空!");
    form1.Author.focus();
    return false;
```

```
            }
        if(Publisher = = "")
        {
            alert("出版社信息不能为空!");
            form1.Publisher.focus();
            return false;
        }
    if(Price = = "")
        {
            alert("单价信息不能为空!");
            form1.Price.focus();
            return false;
        }
        function isNull(str)
        {
            if(str.length = = 0)
            return true;
            else
            return false;
        }

    }


</script>
<html>
<head>
<base href = "<% = basePath%>"/>
<title>413 网上书城|管理员|修改图书</title>
<link href = "29.css" rel = "stylesheet" type = "text/css"/>
    </head>


<body>
<div id = "container">
    <div id = "banner"><img src = "pic1/p01.jpg"/></div>
        <div id = "globallink">
            <ul>
                <li><a href = "enter.jsp">首页</a></li>
                <li><a href = "login.jsp">用户注册</a></li>
                <li><a href = "hot_books.jsp">图书促销</a></li>
                <li><a href = "#">在线阅读</a></li>
                <li><a href = "vip.jsp">会员中心</a></li>
                <li><a href = "about_me.jsp">联系我们</a></li>
                <li><a href = "#">支付方式</a></li>
```

```html
        </ul>
        <br/>
</div>
<div id="left">
    <div id="login">
        网上书店欢迎你  <a href="enter.jsp">[退出]</a>
        <ul>
        <li>图书信息管理
            <ul>
                <li><a href="FindAllBook">查看图书</a></li>
                <li><a href="addbook.jsp">增加图书</a></li>
            </ul>
         </li>
        <li><a href="FindAllOrderServlet">订单管理</a></li>
        <li><a href="FindAllUserServlet">客户管理</a></li>
        <li><a href="LookliuyanServlet">查看留言</a></li>
</ul>
</div>
<div id="category">
    <h4><span>种类</span></h4>
    <ul>
        <li><a href="#">计算机/网络</a></li>
        <li><a href="#">人文科社</a></li>
        <li><a href="#">培训教材</a></li>
        <li><a href="#">经济</a></li>
        <li><a href="#">管理</a></li>
        <li><a href="#">文化/文学</a></li>
        <li><a href="#">百科全书</a></li>
         <li><a href="#">艺术</a></li>
        <li><a href="#">小说</a></li>
        <li><a href="#">期刊杂志</a></li>
    </ul>
    <h4><span>品牌出版社</span></h4>
    <ul>
        <li><a href="#">人民教育出版社</a></li>
        <li><a href="#">现代出版社</a></li>
        <li><a href="#">新世纪出版社</a></li>
        <li><a href="#">中国少儿出版社</a></li>
        <li><a href="#">清华大学出版社</a></li>
        <li><a href="#">泰博文化出版社</a></li>
        <li><a href="#">科学普及出版社</a></li>
    </ul>
```

```html
            </div>
        </div>
        <div id="main2">
        <img src="pic1/latest.jpg"/>
        <img src="pic1/p111.jpg"/>
    <div id="latest">
    <h2 align="center">修改图书信息</h2>
    <form name="form1" action="UpdateBook" method="post" onsubmit="return isValidate(form1)">
        <table align="center">
        <tr>
        <td>图书编号:</td>
        <td><input type="text" name="Book_number" value="${book.book_number}"/></td>
        </tr>
        <tr>
        <td>图书标题:</td>
        <td><input type="text" name="Book_name" value="${book.book_name}"/></td>
        </tr>
        <tr>
        <td>作者:</td>
        <td><input type="text" name="Author" value="${book.author}"/></td>
        </tr>
        <tr>
        <td>出版社:</td>
        <td><input type="text" name="Publisher" value="${book.publisher}"/></td>
        </tr>
        <tr>
        <td>图书价格:</td>
        <td><input type="text" name="Price" value="${book.price}"/></td></tr>
        <tr><td><input type="reset" value="重置"/></td>
            <td><input type="submit" value="提交"/></td>
        </tr>
        </table>
    </form>
    </div>

    <div id="new">
        <ul>
            <li><a href="#"><img src="pic1/3ds max.jpg"/><br/>评论数量:789</a><br/>顾客评分:☆☆☆☆☆</li>
            <li><a href="#"><img src="pic1/SQL Server.jpg"/><br/>评论数量:558</a><br/>顾客评分:☆☆☆☆</li>
            <li><a href="#"><img src="pic1/jsp.jpg"/><br/>评论数量:321</a>
```

```
              <br/>顾客评分:☆☆☆</li>
                         <li><a href="#"><img src="pic1/Linux.jpg"/><br/>评论数量:125</a>
><br/>顾客评分:☆☆</li>
                     </ul>
                         <br/> 
         </div>
         </div>
                     <div id="bottom">
                        <p>413 &copy;版权所有:计软 S09 二班 413 寝室全体成员 <a href="E—mail:
lvzb212@413.com">lvzb212@413.com</a></p>
                    </div>
         </div>
         </body>
</html>
```

3. 编写模型层代码

```
          package bean;

          import java.sql.Connection;
          import java.sql.DriverManager;
          import java.sql.PreparedStatement;
          import java.sql.ResultSet;
          import java.sql.Statement;
          import java.util.ArrayList;

          public class BookBean{
              private String Book_name;
              private String Book_number;
              private float Price;
              private String Author;
              private String Publisher;

              public String getBook_name(){
                  return Book_name;
              }
              public void setBook_name(String book_name){
                  Book_name = book_name;
              }
              public String getBook_number(){
                  return Book_number;
              }
              public void setBook_number(String book_number){
                  Book_number = book_number;
```

```java
    }
    public float getPrice() {
        return Price;
    }
    public void setPrice(float price) {
        Price = price;
    }
    public String getAuthor() {
        return Author;
    }
    public void setAuthor(String author) {
        Author = author;
    }
    public String getPublisher() {
        return Publisher;
    }
    public void setPublisher(String publisher) {
        Publisher = publisher;
    }
    public BookBean()
    {
    }

    public BookBean(String Book_number, String Book_name, String Author, String Publisher,
                    float Price)
    {

        this.Book_number = Book_number;
        this.Book_name = Book_name;
        this.Author = Author;
        this.Publisher = Publisher;
        this.Price = Price;
    }
    public int update() throws Exception
    {

        //连接对象
        Connection con = null;
        Statement stmt = null;
        ResultSet rs = null;
        String url = "jdbc:mysql://localhost:3306/bookstoredb? useUnicode = true&characterEncoding = gb2312";
        String dbuser = "root";
```

```java
            String dbpass = "root";
                //添加语句,根据参数编写动态的添加语句
                    StringBuffer sql = new StringBuffer();
                    sql.append("update t_book set ");
                    sql.append("Book_name = '");
                    sql.append(Book_name);        //标题
                    sql.append("',Author = '");
                    sql.append(Author);           //作者
                    sql.append("',Publisher = '");
                    sql.append(Publisher);        //出版社
                sql.append("',price = '");
                sql.append(Price);                //价格
                sql.append("' where Book_number = '");
                sql.append(Book_number);
                    sql.append("'");
            try
            {
                Class.forName("com.mysql.jdbc.Driver");
                    con = DriverManager.getConnection(url,dbuser,dbpass);
                    stmt = con.createStatement();
                    return stmt.executeUpdate(sql.toString());
            }catch(Exception e)
            {
            }
            finally
            {
                try{ if(rs! = null) rs.close(); }catch(Exception e){}
                try{ stmt.close(); }catch(Exception e){}
                try{ con.close(); }catch(Exception e){}
            }

            return 0;
    }
    public int delete(String Book_number) throws Exception//删除图书信息
        {
            String url = "jdbc:mysql://localhost:3306/bookstoredb? useUnicode = true&characterEncoding = gb2312";
            String dbuser = "root";
            String dbpass = "root";
            // 连接对象
            Connection con = null;
            Statement stmt = null;
            //创建 DBBean 对象
```

```java
        ResultSet rs = null;
    //删除语句
        StringBuffer sql = new StringBuffer();
    sql.append("delete from t_book where Book_number = '");
    sql.append(Book_number);
    sql.append("'");

        try
        {
            //获取连接
            Class.forName("com.mysql.jdbc.Driver");
        con = DriverManager.getConnection(url,dbuser,dbpass);
            stmt = con.createStatement();
    // 执行 sql 语句

            return stmt.executeUpdate(sql.toString());

            //执行删除语句

        }catch(Exception e)
        {
        }
        finally
        {
            try{ if(rs! = null) rs.close(); }catch(Exception e){}
            try{ stmt.close(); }catch(Exception e){}
            try{ con.close(); }catch(Exception e){}
        }

        return 0;
    }

    public void add() throws Exception//添加图书信息
    {// 连接对象
        Connection con = null;
        PreparedStatement stmt = null;
            ResultSet rs = null;

        String url = "jdbc:mysql://localhost:3306/bookstoredb? useUnicode = true&characterEncoding = gb2312";
            String dbuser = "root";
            String dbpass = "root";
    //添加语句,根据参数编写动态的添加语句
```

```java
StringBuffer sql = new StringBuffer();
sql.append("insert into t_book values('");
sql.append(Book_number);// 图书编号
sql.append("','");
sql.append(Book_name);// 书名
sql.append("','");
sql.append(Author);// 作者
sql.append("','");
sql.append(Publisher);// 出版社
sql.append("','");
sql.append(Price);// 价格
sql.append("')");
try
{
    Class.forName("com.mysql.jdbc.Driver");
        con = DriverManager.getConnection(url,dbuser,dbpass);
         stmt = con.prepareStatement(sql.toString());
        stmt.executeUpdate();

}catch(Exception e)
{System.out.println(e.toString());
}
finally
{
    try{ if(rs! = null) rs.close(); }catch(Exception e){}
     try{ stmt.close(); }catch(Exception e){}
     try{ con.close(); }catch(Exception e){}
}}
public boolean hasExist(String Book_number) {
    boolean find = false;
    // 连接对象
    Connection con = null;
    // 结果集对象
    ResultSet rs = null;
    // 创建 DBBean 对象
    PreparedStatement stmt = null;

    // 查询语句
    StringBuffer sql = new StringBuffer();
    sql.append("select * from t_book where Book_number ='");
sql.append(Book_number);

    sql.append("'");
```

```java
        try {
            // 获取连接
            con = DBUtil.getConnection(); // 执行 select 语句,返回结果集对象
            stmt = con.prepareStatement(sql.toString());
            rs = stmt.executeQuery();

            // 指向结果集的第一条
            if (rs.next())
                find = true;
            else
                find = false;
        } catch (Exception e) {
        } finally {
            DBUtil.closeResultSet(rs);
            DBUtil.closeConnection(con);
            DBUtil.closeStatement(stmt);
        }
        // 返回查找的结果
        return find;
    }

    public ArrayList findAllBook(String pageNo) {
        ArrayList booklist = new ArrayList();
        // 连接对象
        Connection con = null;
        // 语句对象
        Statement stmt = null;
        // 结果集对象
        ResultSet rs = null;
        // 查询语句
        String sql = "select * from t_book";
        String url = "jdbc:mysql://localhost:3306/bookstoredb?useUnicode=true&characterEncoding=gbk";
        String dbuser = "root";
        String dbpass = "root";
        try {
            // 创建上下文环境
            Class.forName("com.mysql.jdbc.Driver");
            con = DriverManager.getConnection(url, dbuser, dbpass);
            stmt = con.createStatement();
            rs = stmt.executeQuery(sql);
            // 要显示的页码,默认值为1
            int iPageNo = 1;
```

```java
try {
    // 把字符串转换成整数
    iPageNo = Integer.parseInt(pageNo);
} catch (Exception e) {
}
// 要显示的第一条记录
int begin = (iPageNo - 1) * 3 + 1;
// 要显示的最后一条记录
int end = iPageNo * 3;
// 循环计数器
int index = 0;
// 对结果集进行遍历
while (rs.next()) {
    // 循环到第 index 条
    index++;
    // 如果还没有到要显示的第一条记录,则不处理,继续遍历
    if (index < begin)
        continue;
    // 如果已经大于最后一条记录,则结束循环
    if (index > end)
        break;
    // 获取结果集中的信息
    String Book_number = rs.getString(1);
    String Book_name = rs.getString(2);
    String Author = rs.getString(3);
    String Publisher = rs.getString(4);
    float Price = rs.getFloat(5);

    // 创建图书对象
    BookBean book = new BookBean();
    // 根据获取的图书信息初始化图书对象
    book.setBook_number(Book_number);
    book.setAuthor(Author);
    book.setPublisher(Publisher);
    book.setPrice(Price);
    book.setBook_name(Book_name);
    // 把对象添加到集合中
    booklist.add(book);
}
} catch (Exception e) {
} finally {
    // 关闭对象的顺序:rs stmt con
    // 与创建对象的顺序相反
```

```
                try {
                    rs.close();
                } catch (Exception e) {
                }
                try {
                    stmt.close();
                }  catch (Exception e) {
                }
                try {
                    con.close();
                } catch (Exception e) {
                }
            }
            // 返回查询到的所有对象
            return booklist;
        }
        public BookBean findBookById(String  Book_number)//根据图书的编号查找是否有该图书存在,存在该图书就将该图书的信息返回,返回的是一个对象。
        {
            // 连接对象
            Connection con = null;
            // 结果集对象
            ResultSet rs = null;
            Statement stmt = null;
            String url = "jdbc:mysql://localhost:3306/bookstoredb?useUnicode=true&characterEncoding=gb2312";
            String dbuser = "root";
            String dbpass = "root";

            // 查询语句
            StringBuffer sql = new StringBuffer();
            sql.append("select * from t_book where Book_number = '");
            sql.append(Book_number);
            sql.append("'");
            try {
                // 获取连接
                Class.forName("com.mysql.jdbc.Driver");
                con = DriverManager.getConnection(url, dbuser, dbpass);
                stmt = con.createStatement();
                // 执行 select 语句,返回结果集对象
                rs = stmt.executeQuery(sql.toString());
```

```java
            // 指向结果集的第一条
            if (rs.next()) {

                // 获取结果集中的信息
                Book_number = rs.getString(1);
                String Book_name = rs.getString(2);
                String Author = rs.getString(3);
                String Publisher = rs.getString(4);
                float Price = rs.getFloat(5);

                // 创建图书对象
                BookBean book = new BookBean();

                // 根据获取的图书信息初始化图书对象
                book.setBook_number(Book_number);
                book.setBook_name(Book_name);
                book.setAuthor(Author);
                book.setPublisher(Publisher);
                book.setPrice(Price);
                return book;
            }
        } catch (Exception e) {
            System.out.println(e.toString());
        } finally {

            try{ if(rs! = null) rs.close(); }catch(Exception e){}
            try{ stmt.close(); }catch(Exception e){}
            try{ con.close(); }catch(Exception e){}
        }
        // 返回查找的结果
        return null;
    }
    public Integer getPageCount() {
        int pageCount = 1;
        // 连接对象
        Connection con = null;
        // 语句对象
        Statement stmt = null;
        // 结果集对象
        ResultSet rs = null;
        // 查询语句
        String sql = "select count(*) from t_book";
            String url = " jdbc:mysql://localhost:3306/bookstoredb? useUnicode =
```

```
                    true&characterEncoding = gbk";
                    String dbuser = "root";
                    String dbpass = "root";
                    try {
                        Class.forName("com.mysql.jdbc.Driver");
                        con = DriverManager.getConnection(url, dbuser, dbpass);
                        stmt = con.createStatement();
                        // 执行 select 语句,返回结果集对象
                        rs = stmt.executeQuery(sql);
                        // 指向结果集的第一条
                        rs.next();
                        // 得到第一列
                        int n = rs.getInt(1);
                        // 计算总页数
                        pageCount = (n - 1) / 3 + 1;
                    } catch (Exception e) {
                    } finally {
                        // 关闭对象的顺序:rs stmt con
                        // 与创建对象的顺序相反
                        try {
                            rs.close();
                        } catch (Exception e) {
                        }
                        try {
                            stmt.close();
                        } catch (Exception e) {
                        }
                        try {
                            con.close();
                        } catch (Exception e) {
                        }
                    }
                    // 返回计算的结果
                    return new Integer(pageCount);
                }
        }
```

4. 编写控制层代码

```
        package servlet;

        import java.io.IOException;
        import java.io.PrintWriter;
```

```java
import java.util.ArrayList;

import javax.servlet.RequestDispatcher;
import javax.servlet.ServletException;
import javax.servlet.http.HttpServlet;
import javax.servlet.http.HttpServletRequest;
import javax.servlet.http.HttpServletResponse;
import javax.servlet.http.HttpSession;

import bean.*;

public class FindAllBookServlet extends HttpServlet {

    private static final long serialVersionUID = 1L;

    protected void doGet(HttpServletRequest request, HttpServletResponse response)
            throws ServletException, IOException {

        int pageNo = 1;
        String strpage = request.getParameter("pageNo");
        if(strpage! = null) {
            pageNo = Integer.parseInt(strpage);
        }
        BookBean book = new BookBean();
        try {
            String pageNo2 = String.valueOf(pageNo);
            ArrayList booklist = book.findAllBook(pageNo2);
            request.setAttribute("booklist", booklist);
            Integer pageCount = book.getPageCount();
            request.setAttribute("pageCount", pageCount);
            request.setAttribute("pageNo", pageNo2);
            RequestDispatcher rd = request.getRequestDispatcher("booklist.jsp");
            rd.forward(request, response);
        } catch (Exception e) {
            response.setContentType("text/html;charset=gb2312");
            PrintWriter out = response.getWriter();
            out.println(e.toString());
        }

    }

    protected void doPost(HttpServletRequest request, HttpServletResponse response)
            throws ServletException, IOException {
```

```java
        // TODO Auto-generated method stub
        doGet(request,response);
    }

}


package servlet;

import java.io.*;
import bean.*;
import java.text.*;
import java.util.*;

import javax.servlet.RequestDispatcher;
import javax.servlet.ServletException;
import javax.servlet.http.HttpServlet;
import javax.servlet.http.HttpServletRequest;
import javax.servlet.http.HttpServletResponse;
public class AddBookServlet extends HttpServlet{

    public void doGet(HttpServletRequest request,HttpServletResponse response)
            throws IOException,ServletException
    {
        response.setContentType("text/html;charset=gb2312");
        request.setCharacterEncoding("GB2312");
        //获取用户提交的信息
        String Book_number = request.getParameter("Book_number");
        String Book_name = request.getParameter("Book_name");
        String Author = request.getParameter("Author");
        String Publisher = request.getParameter("Publisher");
        float Price = Float.parseFloat(request.getParameter("Price"));

        //创建JavaBean对象
        BookBean book = new BookBean();
        //初始化
        book.setBook_number(Book_number);
        book.setBook_name(Book_name);
        book.setAuthor(Author);
        book.setPublisher(Publisher);
        book.setPrice(Price);
        //转向的文件
        String forward;
        //提示信息
```

```java
        String info;

            if(book.hasExist(Book_number))
            {
                forward = "addbook.jsp";
                info = "该图书编号已经存在!";
            }
            else
            {try{
                book.add();
                info = "添加图书成功!";
                forward = "FindAllBook";}

            catch(Exception e){
                forward = "addbook.jsp";
                info = "系统异常!";}
        }

            request.setAttribute("info",info);

            //定义跳转文件
            RequestDispatcher rd = request.getRequestDispatcher(forward);
            //完成重定向
            rd.forward(request,response);
        }

    public void doPost(HttpServletRequest request,HttpServletResponse response)
        throws IOException,ServletException
    {
        doGet(request,response);
    }

}


package servlet;
import javax.servlet.*;
import javax.servlet.http.*;
import java.io.*;
import bean.*;

public class DeleteBookServlet extends HttpServlet
{
    private static final long serialVersionUID = 1L;
```

```java
public void doGet(HttpServletRequest request, HttpServletResponse response)
        throws IOException, ServletException {
    //获取用户提交的信息
    String Book_number = request.getParameter("Book_number");

    //创建JavaBean对象
    BookBean book = new BookBean();

    //提示信息
    String info = null;
    try{
        if(book.delete(Book_number) > 0)
        {
            info = "成功删除图书信息!";
        }else{
            info = "删除图书信息失败!";
        }
    }catch(Exception e){
        info = "数据库异常!";
    }

    request.setAttribute("info", info);
    //定义跳转文件
    RequestDispatcher rd = request.getRequestDispatcher("FindAllBook");
    //完成重定向
    rd.forward(request, response);
}
public void doPost(HttpServletRequest request, HttpServletResponse response)
        throws IOException, ServletException {
    doGet(request, response);
}
}


package servlet;
import javax.servlet.*;
import javax.servlet.http.*;
import java.io.*;
import bean.*;

public class UpdateBookServlet extends HttpServlet
{ private static final long serialVersionUID = 1L;
public void doGet(HttpServletRequest request, HttpServletResponse response)
```

```java
    throws IOException,ServletException  {
        //获取用户提交的信息
response.setContentType("text/html;charset=gb2312");
request.setCharacterEncoding("GB2312") ;

    String Book_number = request.getParameter("Book_number");
    String Book_name = request.getParameter("Book_name");
    String Author = request.getParameter("Author");
    String Publisher = request.getParameter("Publisher");
    float Price = Float.parseFloat(request.getParameter("Price"));

    //创建 JavaBean 对象
    BookBean book = new BookBean();

    //初始化
    book.setBook_number(Book_number);
    book.setAuthor(Author);
    book.setPublisher(Publisher);
    book.setPrice(Price);
    book.setBook_name(Book_name);

    //提示信息
    String info;
    try{
        if(book.update()>0){
            info = "图书信息修改成功!";
        }else{
            info = "图书信息修改失败!";
        }
    }catch(Exception e){
        info = "数据库异常!";
    }
            request.setAttribute("info",info);
    //定义跳转文件
    RequestDispatcher rd = request.getRequestDispatcher("FindAllBook");
    //完成重定向
    rd.forward(request,response);
}
public void doPost(HttpServletRequest request,HttpServletResponse response)
    throws IOException,ServletException  {
    doGet(request,response);
}
}
```

```java
package servlet;
import javax.servlet.*;
import bean.*;

import java.io.IOException;
import javax.servlet.ServletException;
import javax.servlet.http.HttpServlet;
import javax.servlet.http.HttpServletRequest;
import javax.servlet.http.HttpServletResponse;
import javax.servlet.http.HttpSession;

public class UpdateFindBookServlet extends HttpServlet
{ private static final long serialVersionUID = 1L;
public void doGet(HttpServletRequest request,HttpServletResponse response)
      throws IOException,ServletException  {

//获取用户提交的信息
String Book_number = request.getParameter("Book_number");

//创建 JavaBean 对象
BookBean ub = new BookBean();
//转向的文件
String forward;
//提示信息
String info = null;
BookBean book = ub.findBookById(Book_number);
if(book  = =  null)
{
    forward = "FindAllBook";
    info = "要修改的用户不存在!";
    //保存提示信息
    request.setAttribute("info",info);
}else{
    request.setAttribute("book",book);
    forward = "updatebook.jsp";
}
//定义跳转文件
RequestDispatcher rd = request.getRequestDispatcher(forward);
//完成重定向
rd.forward(request,response);
}
public void doPost(HttpServletRequest request,HttpServletResponse response)
```

```
        throws IOException, ServletException {
            doGet(request, response);
        }
}
```

任务十一：Blog 系统日志信息模块

一、任务描述

你作为《Blog 系统》项目开发组的程序员，请实现如下功能：
- 日志信息的添加、编辑和删除；
- 日志信息的列表显示。

二、功能描述

日志管理子模块的页面如图 11.1、图 11.2 所示。

图 11.1　日志管理子模块页面

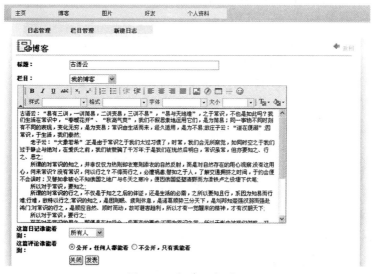

图 11.2　新建日志页面

三、要求

1. 数据库实现

（1）新建数据库，数据库名：BlogDB。

（2）在 BlogDB 数据库中新建数据表：用户日志表，数据表名：T_user_log，见表 11.1。

表 11.1 用户日志表（T_user_log）表结构

字段名	字段说明	字段类型	是否允许为空	备注
User ＜pk＞	用户名	varchar(30)	否	
Title	日志名称	varchar(60)	否	
Columns	日志栏目	varchar(60)	否	
Content	日志内容	text	否	
Permissions	日志权限	varchar(10)	否	
Comment	评论	text	是	
Published_time	发表时间	timestamp	否	当前系统时间
Set_top	是否置顶	boolean	否	

2. 功能实现

在栏目"博客"中的"日志管理"页面中实现如下功能：

（1）点击框架左边导航条中的"日志管理"，在右边的主体部分中显示日志管理信息列表，如图 11.1 所示。

（2）点击日志信息列表页面上的"写博客"或者"新建日志"按钮，进入"日志录入"页面，如图 11.2 所示。

（3）点击日志录入页面中的"发表"按钮，在数据库的 T_user_log 表中增加一条日志信息，增加前对必填项进行判断。

（4）点击日志信息列表页面中某信息的"编辑"按钮，实现该日志的编辑功能。

（5）点击日志信息列表页面中某信息的"删除"按钮，实现该日志的删除功能。

（6）日志信息增加或删除成功后，自动定位到日志信息列表页面，显示更新后的日志信息列表，如图 11.1 所示。

（7）测试程序，增加两条以上日志信息进入数据库。

四、必备知识

1. 数据库相关知识

（1）使用 MS SQL Server 2005/2008 创建数据库，创建数据表，设置表的字段，数据类型，主键，外键，约束。

（2）向数据表插入、删除、修改、查询数据。

2. 页面相关知识

(1)使用 HTML 制作项目页面。

(2)使用 CSS 控制页面的样式。

(3)使用 JavaScript 对页面必要的内容进行校验。

3. JSP 相关知识

(1)使用 JSTL 标准标签库控制页面显示逻辑。

(2)理解 JSP 的 request,response,session,application 的概念。

(3)使用 EL 表达式在页面显示数据。

(4)合理地使用转发和重定向控制项目的页面跳转。

(5)使用 JDBC 与数据库进行交互。

(6)MVC 模式下的分层架构,控制器,视图,模型的划分和通信。

五、解题思路

1. 数据库思路

(1)根据项目要求创建数据库和数据表,向数据表中插入合适的测试数据。

(2)导入 JDBC 驱动包,编写 JDBC 的连接工具代码。

(3)编写数据操作对象代码,负责进行与数据库的交互操作。

2. 视图层思路

(1)将提供的素材页面改写为 JSP 页面。

(2)JSP 使用 JSTL 和 EL 负责控制页面显示的逻辑。

3. 控制层思路

(1)使用 Servlet 类控制一次请求响应过程的处理。

(2)由 Servlet 按照顺序进行请求的处理、数据库交互、模型存取和封装、页面跳转逻辑控制等。

4. 模型层思路

使用 JavaBean 作为模型层,封装数据和行为。

六、操作步骤

1. 准备数据库

(1)根据项目要求,在 SQL Server2008 中创建 BlogDB 数据库、用户日志表(T_user_log),并插入测试数据。

(2)编写 JDBC 的连接工具代码。

```
package blog.database;

import java.sql.Connection;
```

```java
import java.sql.DriverManager;
import java.sql.PreparedStatement;
import java.sql.ResultSet;
import java.sql.SQLException;

public class BlogConnection {

    public static Connection getConnection() {
        Connection connection = null;
        try {
            Class.forName("com.mysql.jdbc.Driver");
            connection = DriverManager.getConnection("jdbc:mysql://localhost:3306/blogdb? useUnicode = true&characterEncoding = gb2312", "root", "root");
        } catch (ClassNotFoundException e) {

            e.printStackTrace();
        } catch (SQLException e) {
            e.printStackTrace();
        }

        return connection;
    }

    public static PreparedStatement getParameter(Connection connection, String sql) {
        PreparedStatement pStatement = null;
        try {
            pStatement = connection.prepareStatement(sql);

        } catch (SQLException e) {

            e.printStackTrace();
        }

        return pStatement;

    }

    public static ResultSet getResultSet(PreparedStatement pStatement) {
        ResultSet rSet = null;

        try {
            rSet = pStatement.executeQuery();
        } catch (SQLException e) {
```

```java
                e.printStackTrace();
            }
            return rSet;
        }

        public static void closing(Connection connection){
            try{
                if(connection! = null){
                    connection.close();
                    connection = null;
                }

            }catch(SQLException e){

                e.printStackTrace();
            }

        }

        public static void closing(PreparedStatement pStatement){

            try{
                if(pStatement! = null){
                    pStatement.close();
                    pStatement = null;
                }

            }catch(SQLException e){

                e.printStackTrace();
            }

        }

        public static void closing(ResultSet rSet){

            try{
                if(rSet! = null){
                    rSet.close();
                    rSet = null;
                }
```

```java
        } catch (SQLException e) {
            e.printStackTrace();
        }
    }
}

package blog.database.user;

import java.sql.Connection;
import java.sql.PreparedStatement;
import java.sql.ResultSet;
import java.sql.SQLException;
import java.util.ArrayList;
import java.util.List;

import blog.bean.user.Log;
import blog.database.BlogConnection;
import blog.inter.user.ILog;

public class UserLog implements ILog {
    private Connection connection;
    public UserLog(Connection connection) {
        this.connection = connection;

    }
    //删除  按User来删除
    public void delete(String user) {
        PreparedStatement pStatement = null;
        String sql;
        sql = "delete from t_user_log where user = ?";
        try {
            pStatement = BlogConnection.getParameter(connection, sql);
            pStatement.setString(1, user);
            pStatement.executeUpdate();

        } catch (SQLException e) {

            e.printStackTrace();
        }
        BlogConnection.closing(pStatement);
```

```java
}
//插入
public void insert(Log log) {
    PreparedStatement pStatement;
    String sql;
    sql = " insert into t_user_log ( user, title, columns, content, permissions, comment ) values(?,?,?,?,?,?)";
    pStatement = BlogConnection.getParameter(connection, sql);
    try {
        pStatement.setString(1, log.getUser());
        pStatement.setString(2, log.getTitle());
        pStatement.setString(3, log.getColumns());
        pStatement.setString(4, log.getContent());
        pStatement.setString(5, log.getPermissions());
        pStatement.setString(6, log.getComment());
        pStatement.executeUpdate();
        BlogConnection.closing(pStatement);
    } catch (SQLException e) {

        e.printStackTrace();
    }

}
//更新
public void update(Log log) {
    PreparedStatement pStatement;
    String sql;
    sql = " update t_user_log set title = ?, columns = ?, content = ?, permissions = ?, comment = ? where user = ?";
    pStatement = BlogConnection.getParameter(connection, sql);

    try {

        pStatement.setString(1, log.getTitle());
        pStatement.setString(2, log.getColumns());
        pStatement.setString(3, log.getContent());
        pStatement.setString(4, log.getPermissions());
        pStatement.setString(5, log.getComment());

        pStatement.setString(6, log.getUser());
        pStatement.executeUpdate();
        BlogConnection.closing(pStatement);
    } catch (SQLException e) {
```

```java
        e.printStackTrace();
    }

}

//查看
public List<Log> allfind(String user, int pageN, int pageLimit, String columns) {
    List<Log> list = new ArrayList<Log>();
    PreparedStatement pStatement;
    ResultSet rSet = null;
    String sql;

    sql = "select * from t_user_log where columns = ? order by user desc limit ?,? ";

    pStatement = BlogConnection.getParameter(connection, sql);

    try {

        pStatement.setString(1, columns);
        pStatement.setInt(2, pageN);
        pStatement.setInt(3, pageLimit);

        rSet = BlogConnection.getResultSet(pStatement);
        while (rSet.next()) {
            Log log = new Log();
            log.setColumns(rSet.getString("columns"));
            log.setComment(rSet.getString("comment"));
            log.setContent(rSet.getString("content"));
            log.setPermissions(rSet.getString("permissions"));
            log.setPublishedTime(rSet.getString("publishedTime"));
            log.setTitle(rSet.getString("title"));
            log.setUser(rSet.getString("user"));
            log.setSettop(rSet.getString("settop"));
            list.add(log);
        }

    } catch (SQLException e) {

        e.printStackTrace();
    }
    BlogConnection.closing(rSet);
    BlogConnection.closing(pStatement);
    return list;
```

}

```java
public Log logfind(String user) {

    Log log = null;
    PreparedStatement pStatement;
    ResultSet rSet = null;
    String sql;
    sql = "select * from t_user_log where user = ?";
    pStatement = BlogConnection.getParameter(connection, sql);

    try {

        pStatement.setString(1, user);
        rSet = BlogConnection.getResultSet(pStatement);
        if (rSet.next()) {
            log = new Log();
            log.setColumns(rSet.getString("columns"));
            log.setComment(rSet.getString("comment"));
            log.setContent(rSet.getString("content"));

            log.setPermissions(rSet.getString("permissions"));
            log.setPublishedTime(rSet.getString("publishedTime"));
            log.setTitle(rSet.getString("title"));
            log.setUser(rSet.getString("user"));

        }

    } catch (SQLException e) {

        e.printStackTrace();
    }
    BlogConnection.closing(rSet);
    BlogConnection.closing(pStatement);
    return log;
}
```

}

package blog.database.user.proxy;

import java.sql.Connection;

```java
import java.util.List;

import org.omg.CORBA.PUBLIC_MEMBER;

import blog.bean.user.Log;

import blog.database.BlogConnection;
import blog.database.user.UserLog;
import blog.inter.user.ILog;

public class UserLogProxy implements ILog {
    private Connection connection;
    private UserLog uLog;
    public UserLogProxy() {
        this.connection = BlogConnection.getConnection();
        uLog = new UserLog(this.connection);
    }
    public List<Log> allfind(String user, int pageN, int pageLimit, String columns) {
        List<Log> list = null;
        try {
            list = uLog.allfind(user, pageN, pageLimit, columns);
        } catch (Exception e) {
            e.printStackTrace();
        } finally {
            BlogConnection.closing(connection);
        }
        return list;
    }

    public void delete(String user) {
        try {
            if (uLog.logfind(user) != null) {
                uLog.delete(user);
            }

        } catch (Exception e) {
            e.printStackTrace();
        } finally {
            BlogConnection.closing(connection);
        }
    }

    public void insert(Log log) {
```

```java
        try {
            if (uLog.logfind(log.getUser()) == null) {
                uLog.insert(log);
            }
        } catch (Exception e) {
            e.printStackTrace();
        } finally {
            BlogConnection.closing(connection);
        }

    }

    public Log logfind(String user) {
        Log log = null;
        try {
            if (uLog.logfind(user) != null) {
                log = uLog.logfind(user);
            }
        } catch (Exception e) {
            e.printStackTrace();
        } finally {
            BlogConnection.closing(connection);
        }
        return log;
    }

    public void update(Log log) {
        try {
            if (uLog.logfind(log.getUser()) != null) {
                uLog.update(log);
            }
        } catch (Exception e) {
            e.printStackTrace();
        } finally {
            BlogConnection.closing(connection);
        }
    }

}

package blog.Factory;
```

```java
import blog.database.user.proxy.UserLogProxy;
public class DAOFactory{
    public static UserLogProxy getUserLogProxy(){
        return new UserLogProxy();
    }
}
```

2. 编写视图层代码

```jsp
<%@ page language="java" import="java.util.*" pageEncoding="UTF-8"%>
<%@ taglib prefix="c" uri="http://java.sun.com/jsp/jstl/core" %>
<%@ taglib prefix="fn" uri="http://java.sun.com/jsp/jstl/functions" %>
<!DOCTYPE html PUBLIC "-//W3C//DTD XHTML 1.0 Transitional//EN" "http://www.w3.org/TR/xhtml1/DTD/xhtml1-transitional.dtd">
<html xmlns="http://www.w3.org/1999/xhtml">
<head>
<meta http-equiv="Content-Type" content="text/html; charset=utf-8" />
<title>css web layout by www.865171.cn</title>
<link rel="stylesheet" href="assets/css/style.css" type="text/css" media="all" />
<script src="SpryAssets/SpryMenuBar.js" type="text/javascript"></script>
<link href="SpryAssets/SpryMenuBarHorizontal.css" rel="stylesheet" type="text/css" />
<link href="SpryAssets/SpryMenuBarVertical.css" rel="stylesheet" type="text/css" />
</head>
<body>
<div id="header">
  <p> </p>
  <p> </p>
  <p> </p>
  <p> </p>
  <p> </p>
  <p> </p>
  <p> </p>
  <p> </p>
  <ul id="MenuBar1" class="MenuBarHorizontal">
    <li><a href="LogManagement">主页</a>    </li>
    <li><a href="LogManagement">博客</a></li>
    <li><a href="LogManagement">图片</a>    </li>
    <li><a href="LogManagement">好友</a></li>
    <li><a href="LogManagement">个人资料</a></li>
  </ul>
  <p> </p>
</div>
```

```html
<div id="contentwrap">
    <div id="breadcrumb">
        <ul id="MenuBar4" class="MenuBarHorizontal">
            <li><a href="LogManagement">日志管理</a></li>
            <li><a href="LogManagement">栏目管理</a></li>
            <li><a href="user_newlog.jsp">新建日志</a></li>
        </ul>
        <p> </p>
        <p> </p>
    </div>
    <div id="content">
        <table width="318" height="24" border="0">
            <tr>
                <td width="153" height="18"><img src="admin/images/report2_(add).gif" width="34" height="26"/><font size="+1">博客</font></td>
                <td width="149"><a href="user_newlog.jsp"><img src="admin/images/edit.gif" width="32" height="24" style="border-color:#FFF"/><font size="+1">写博客</font></a></td>
            </tr>
        </table><hr/>
        <div id="bu">

            <p><a href="LogManagement">我的博客</a></p>

            <p> </p>
            <p> </p>
            <p> </p>
            <p> </p>
            <p> </p>
            <p> </p>
            <p> </p>
            <p> </p>
            <p> </p>
            <p> </p>
            <p> </p>
        </div>
        <div id="bu2">
            <table width="100%" height="99" align="right" cellpadding="1" cellspacing="1">
                <tr bgcolor="#eff0e0" align="center">
                    <td width="182" height="26"><strong>题目</strong></td>
                    <td width="99"><strong>栏目</strong></td>
                    <td width="109"><strong>发布日期</strong></td>
```

```
            <td width="55"><strong>状态</strong></td>
            <td width="55"><strong>置顶</strong></td>
            <td width="37"><strong>编辑</strong></td>
            <td width="39"><strong>删除</strong></td>
        </tr>
        <c:forEach items="${logs}" var="lo">
        <tr bgcolor="#FFFFFF" align="center">
            <td width="182" align="left" style="border-bottom-color:#999">${lo.title}</td>
            <td width="99" align="center" style="border-bottom-color:#999">
```

我的博客

```
            </td>
            <td align="center" style="border-bottom-color:#999">${fn:substring(lo.publishedTime,0,16)}</td>
            <td width="55" align="center">
                <c:if test="${0==lo.permissions}">公开</c:if>
                <c:if test="${1==lo.permissions}">仅好友</c:if>
                <c:if test="${2==lo.permissions}">隐藏</c:if>
            </td>
            <td align="center" style="border-bottom-color:#999"><c:if test="${0==lo.settop}">置顶</c:if>
                <c:if test="${1==lo.settop}">取消置顶</c:if></td>
            <td align="center" style="border-bottom-color:#999"><a href="EditorLog?user=${lo.user}"><img src="admin/images/e1.jpg" alt="编辑" width="20" height="18" style="border-color:#FFF"/></a></td>
            <td align="center" style="border-bottom-color:#999"><a href="DeleteLog?user=${lo.user}" onclick="return confirm('是否确认删除这篇《${lo.title}》日志');"><img src="admin/images/close.gif" width="20" height="18" style="border-color:#FFF" alt="删除"/></a></td>
        </tr>
        <tr>
            <td align="center" colspan="7"><hr/></td>
        </tr>
        </c:forEach>

        <tr><td align="center" colspan="7"><table cellpadding="0" cellspacing="0">
            <tr>
```

```
            <td align="center">
            </td> </tr>
        </table>
        <p> </p>

        <p> </p>
        <p> </p>
        <p> </p>
        <p> </p>

        <p> </p>
        <p> </p>
        <p> </p>
        <p> </p>
        <p> </p>
        <p> </p>
        <p> </p>
        <p> </p>
        <p> </p>
        <p> </p>
        <p> </p>
        <p> </p>

        <p> </p>
        <div id="footer" class="clear"> <img src="assets/images/logo_footer.gif" width="55" height="54" class="floatleft" />
            <p> </p>
            <p> </p>
            <p> </p>
        </div>

    <script type="text/javascript">
    <!--
    var MenuBar1 = new Spry.Widget.MenuBar("MenuBar1", {imgDown:"SpryAssets/SpryMenuBarDownHover.gif", imgRight:"SpryAssets/SpryMenuBarRightHover.gif"});
    var MenuBar2 = new Spry.Widget.MenuBar("MenuBar2", {imgRight:"SpryAssets/SpryMenuBarRightHover.gif"});
    var MenuBar3 = new Spry.Widget.MenuBar("MenuBar3", {imgRight:"SpryAssets/SpryMenuBarRightHover.gif"});
    var MenuBar4 = new Spry.Widget.MenuBar("MenuBar4", {imgDown:"SpryAssets/SpryMenuBarDownHover.gif", imgRight:"SpryAssets/SpryMenuBarRightHover.gif"});
    //-->
```

```jsp
        </script>
     </body>
</html>

        <%@ page language="java" import="java.util.*" pageEncoding="UTF-8"%>
        <%@ taglib prefix="c" uri="http://java.sun.com/jsp/jstl/core" %>
        <%@ taglib prefix="fn" uri="http://java.sun.com/jsp/jstl/functions" %>
        <%
     String path = request.getContextPath();
     String basePath = request.getScheme()+"://"+request.getServerName()+":"+request.getServerPort()+path+"/";
        %>

        <!DOCTYPE html PUBLIC "-//W3C//DTD XHTML 1.0 Transitional//EN" "http://www.w3.org/TR/xhtml1/DTD/xhtml1-transitional.dtd">
        <html xmlns="http://www.w3.org/1999/xhtml">
        <head>
        <meta http-equiv="Content-Type" content="text/html; charset=utf-8" />
        <title>文本区域</title>
        <link rel="stylesheet" href="assets/css/style.css" type="text/css" media="all" />
        <script src="SpryAssets/SpryMenuBar.js" type="text/javascript"></script>
        <link href="SpryAssets/SpryMenuBarHorizontal.css" rel="stylesheet" type="text/css" />
        <link href="SpryAssets/SpryMenuBarVertical.css" rel="stylesheet" type="text/css" />
        <script type="text/javascript" src="fckeditor/fckeditor.js"></script>
        <script Charset="gb2312" Type="Text/JavaScript" language="javascript" src="js/user_log.js"></script>
          <script type="text/javascript">

     window.onload = function()
     {
            var sBasePath = "<%=request.getScheme()+"://"+request.getServerName()+":"+request.getServerPort()+request.getContextPath()+"/fckeditor/"%>";

            var oFCKeditor = new FCKeditor('content');
            oFCKeditor.Width='90%';
            oFCKeditor.Height='300';
            oFCKeditor.BasePath = sBasePath;
            oFCKeditor.ReplaceTextarea();
     }
     function Myopen(divID){

            var div = document.getElementById(divID);
            div.style.display='block';
```

```
            div.style.left=(document.body.clientWidth-400)/2;
            div.style.top=(document.body.clientHeight-1050)/2;
        }

        function Myclose(divID){
            document.getElementById(divID).style.display='none';
        }

    </script>
</head>
<body>
<div id="header">
    <p> </p>
    <p> </p>
    <p> </p>
    <p> </p>
    <p> </p>
    <p> </p>
    <p> </p>
    <p> </p>
    <p> </p>
    <ul id="MenuBar1" class="MenuBarHorizontal">
        <li><a href="LogManagement">主页</a>    </li>
        <li><a href="LogManagement">博客</a></li>
        <li><a href="LogManagement">图片</a>    </li>
        <li><a href="LogManagement">好友</a></li>
        <li><a href="LogManagement">个人资料</a></li>
    </ul>
    <p> </p>
</div>
<div id="contentwrap">
    <div id="breadcrumb">
        <ul id="MenuBar4" class="MenuBarHorizontal">
            <li><a href="LogManagement">日志管理</a></li>
            <li><a href="LogManagement">栏目管理</a></li>
            <li><a href="user_newlog.jsp">新建日志</a></li>
        </ul>
        <p> </p>
        <p> </p>
    </div>
    <div id="content">
        <table width="756" height="24" border="0">
            <tr>
```

```
            <td width="658" height="18"><img src="admin/images/report2_(add).gif" width="34" height="26"/><font size="+1">博客</font></td>
            <td width="88" align="right"><a href="LogManagement?user=${user}"><img src="admin/images/pic25.gif" width="20" height="14" style="border-color:#FFF"/><font color="#00FF33"><b>返回</b></font></a></td>
          </tr>
        </table>

        <hr/>
        <form id="form1" name="form1" method="post" action="EditorLog?user=${user}&add=1" onsubmit="replychange()">
          <table width="674" height="348" border="0">
            <tr>
              <td width="113" height="34"><span class="td1">标题：</span></td>
              <td width="543"><label>
                <input name="title" type="text" id="title" size="50" maxlength="30" value="${log.title}"/>
              </label></td>
            </tr>
            <tr>
              <td height="21">栏目：</td>
              <td><label>
                <select name="columns" id="columns">
                  <option value="1" selected="selected">我的博客</option>
                </select>
              </label></td>
            </tr>
            <tr>
              <td height="81" colspan="2"><label>
                <textarea name="content" cols="80" rows="5" id="content">${log.content}</textarea>
              </label></td>
            </tr>
            <tr>
              <td>这篇日记谁能看到：</td>
              <td><label>
                <select name="permissions" id="permissions">
                  <c:choose>
                    <c:when test="${log.permissions==0}">
                      <option value="0" selected="selected">所有人</option>
```

```
              <option value="1">好友</option>
              <option value="2">完全隐藏</option>
            </c:when>
            <c:when test="${log.permissions==1}">
              <option value="0">所有人</option>
              <option value="1" selected="selected">好友</option>
              <option value="2">完全隐藏</option>
            </c:when>
            <c:otherwise>
              <option value="0">所有人</option>
              <option value="1">好友</option>
              <option value="2" selected="selected">完全隐藏</option>
            </c:otherwise>
          </c:choose>

        </select>
      </label></td>
</tr>
<tr>
  <td>这篇评论谁能看到：</td>
  <td><span id="f_Diary_ShowType">
    <c:choose>
      <c:when test="${log.comment==0}">
        <input type="radio" checked="checked" value="0" name="comment" />
        公开,任何人都能看
        <input type="radio" value="1" name="comment" />
        不公开,只有我能看
      </c:when>
      <c:otherwise>
        <input type="radio" value="0" name="comment" />
        公开,任何人都能看
        <input type="radio" value="1" name="comment" checked="checked" />
        不公开,只有我能看
      </c:otherwise>
    </c:choose>
    </span></td>
</tr>
<tr>
  <td> </td>
  <td><label>
    <input type="reset" name="button" id="button" value="关闭" />
  </label>
    <label>
```

```html
            <input name="button2" type="submit" id="button2" value="发表"/>
          </label></td>
        </tr>
      </table>
      <p> </p>
      <p> </p>
      <p> </p>
    </form>
    <p> </p>
    <p> </p>
    <p> </p>
    <p> </p>
    <p> </p>
    <p> </p>
    <p> </p>
    <p> </p>
    <p> </p>
    <p> </p>
    <p> </p>
    <p> </p>
  </div>
  <p> </p>
  <div id="footer" class="clear"> <img src="assets/images/logo_footer.gif" width="55" height="54" class="floatleft"/>
    <p> </p>
    <p> </p>
    <p> </p>
  </div>
</div>
<script type="text/javascript">
<!--
var MenuBar1 = new Spry.Widget.MenuBar("MenuBar1", {imgDown:"SpryAssets/SpryMenuBarDownHover.gif", imgRight:"SpryAssets/SpryMenuBarRightHover.gif"});
var MenuBar2 = new Spry.Widget.MenuBar("MenuBar2", {imgRight:"SpryAssets/SpryMenuBarRightHover.gif"});
var MenuBar3 = new Spry.Widget.MenuBar("MenuBar3", {imgRight:"SpryAssets/SpryMenuBarRightHover.gif"});
var MenuBar4 = new Spry.Widget.MenuBar("MenuBar4", {imgDown:"SpryAssets/SpryMenuBarDownHover.gif", imgRight:"SpryAssets/SpryMenuBarRightHover.gif"});
```

```jsp
//-->
</script>
</body>
</html>

<%@ page language="java" import="java.util.*" pageEncoding="UTF-8"%>
<%@ taglib prefix="c" uri="http://java.sun.com/jsp/jstl/core" %>
<%@ taglib prefix="fn" uri="http://java.sun.com/jsp/jstl/functions" %>
<%
String path = request.getContextPath();
String basePath = request.getScheme()+"://"+request.getServerName()+":"+request.getServerPort()+path+"/";
%>

<!DOCTYPE html PUBLIC "-//W3C//DTD XHTML 1.0 Transitional//EN" "http://www.w3.org/TR/xhtml1/DTD/xhtml1-transitional.dtd">
<html xmlns="http://www.w3.org/1999/xhtml">
<head>
<meta http-equiv="Content-Type" content="text/html; charset=utf-8" />
<title>文本区域</title>
<link rel="stylesheet" href="assets/css/style.css" type="text/css" media="all" />
<script src="SpryAssets/SpryMenuBar.js" type="text/javascript"></script>
<link href="SpryAssets/SpryMenuBarHorizontal.css" rel="stylesheet" type="text/css" />
<link href="SpryAssets/SpryMenuBarVertical.css" rel="stylesheet" type="text/css" />
<script type="text/javascript" src="fckeditor/fckeditor.js"></script>
<script Charset="gb2312" Type="Text/JavaScript" language="javascript" src="js/user_log.js"></script>
<script language="javascript">
function init1(){
    alert("${error}");
}


window.onload=function()
{
    <c:if test="${!empty error}">
    init1();
    </c:if>
    var sBasePath = "<%=request.getScheme()+"://"+request.getServerName()+":"+request.getServerPort()+request.getContextPath()+"/fckeditor/"%>";

    var oFCKeditor = new FCKeditor('content') ;
    oFCKeditor.Width = '90%';
```

```
        oFCKeditor.Height = '300';
        oFCKeditor.BasePath = sBasePath;
        oFCKeditor.ReplaceTextarea();
    init();
}
function Myopen(divID){
    var div = document.getElementById(divID);
    div.style.display = 'block';
    div.style.left = (document.body.clientWidth - 400)/2;
    div.style.top = (document.body.clientHeight - 105)/2;
}

function Myclose(divID){
    document.getElementById(divID).style.display = 'none';
}

drag = 0
move = 0
function init(){
    window.document.onmousemove = mouseMove
    window.document.onmousedown = mouseDown
    window.document.onmouseup = mouseUp
    window.document.ondragstart = mouseStop
}
function mouseDown(){
    if (drag){
        clickleft = window.event.x - parseInt(dragObj.style.left)
        clicktop = window.event.y - parseInt(dragObj.style.top)
        dragObj.style.zIndex += 1
        move = 1
    }
}

function mouseStop(){
    window.event.returnValue = false
}

function mouseMove(){
    if (move){
        dragObj.style.left = window.event.x - clickleft
        dragObj.style.top = window.event.y - clicktop
    }
}
```

```
            }

            function mouseUp() {
                move = 0
            }

        </script>
    </head>
    <body>
    <div id = "header">
        <p> </p>
        <p> </p>
        <p> </p>
        <p> </p>
        <p> </p>
        <p> </p>
        <p> </p>
        <p> </p>
        <p> </p>
        <ul id = "MenuBar1" class = "MenuBarHorizontal">
            <li><a href = "LogManagement">主页</a>    </li>
            <li><a href = "LogManagement">博客</a></li>
            <li><a href = "LogManagement">图片</a>    </li>
            <li><a href = "LogManagement">好友</a></li>
            <li><a href = "LogManagement">个人资料</a></li>
        </ul>
        <p> </p>
    </div>
    <div id = "contentwrap">
        <div id = "breadcrumb">
            <ul id = "MenuBar4" class = "MenuBarHorizontal">
                <li><a href = "LogManagement">日志管理</a></li>
                <li><a href = "LogManagement">栏目管理</a></li>
                <li><a href = "user_newlog.jsp">新建日志</a></li>
            </ul>
            <p> </p>
            <p> </p>
        </div>
        <div id = "content">
            <table width = "756" height = "24" border = "0">
                <tr>
                    <td width = "658" height = "18"><img  src = "admin/images/report2_(add).gif" width = "34" height = "26"/><font size = " +1">博客</font></td>
```

```html
            <td width="88" align="right"><a href="LogManagement?user=${user}"><img src="admin/images/pic25.gif" width="20" height="14" style="border-color:#FFF"/><font color="#00FF33"><b>返回</b></font></a></td>
          </tr>
        </table>
        <hr/>
        <form id="form1" name="form1" method="post" action="NewLog" onsubmit="return replychange()">
            <table width="674" height="348" border="0">
              <tr>
                <td width="113" height="34"><span class="td1">标题：</span></td>
                <td width="543"><label>
                    <input name="title" type="text" id="title" size="50" maxlength="30"/>
                </label></td>
              </tr>
              <tr>
                <td height="21">栏目：</td>
                <td><label>
                    <select name="columns" id="columns">

                        <option value="1">我的博客</option>

                    </select>
                </label></td>
              </tr>
              <tr>
                <td height="81" colspan="2"><label>
                    <textarea name="content" cols="80" rows="5" id="content"></textarea>
                </label></td>
              </tr>
              <tr>
                <td>这篇日记谁能看到：</td>
                <td><label>
                    <select name="permissions" id="permissions">
                        <option value="0" selected="selected">所有人</option>
                        <option value="1">好友</option>
                        <option value="2">完全隐藏</option>
                    </select>
                </label></td>
              </tr>
              <tr>
                <td>这篇评论谁能看到：</td>
                <td><span id="f_Diary_ShowType">
```

```
            <input type = "radio" checked = "checked" value = "0" name = "comment" />
            公开,任何人都能看
            <input type = "radio" value = "1" name = "comment" />
            不公开,只有我能看 </span> </td>
          </tr>
          <tr>
            <td> </td>
            <td><label>
              <input type = "reset" name = "button" id = "button" value = "关闭" />
            </label>
              <label>
                <input name = "button2" type = "submit" id = "button2" value = "发表" />
              </label></td>
          </tr>
        </table>
        <p> </p>
      </form>
      <p> </p>
    </div>
    <p> </p>
    <div id = "footer" class = "clear">  <img src = "assets/images/logo_footer.gif" width = "55" height = "54" class = "floatleft" />
      <p> </p>
      <p> </p>
      <p> </p>
    </div>
</div>


<script type = "text/javascript">
<!--
var MenuBar1 = new Spry.Widget.MenuBar("MenuBar1", {imgDown:"SpryAssets/SpryMenuBarDownHover.gif", imgRight:"SpryAssets/SpryMenuBarRightHover.gif"});
var MenuBar2 = new Spry.Widget.MenuBar("MenuBar2", {imgRight:"SpryAssets/SpryMenuBarRightHover.gif"});
var MenuBar3 = new Spry.Widget.MenuBar("MenuBar3", {imgRight:"SpryAssets/SpryMenuBarRightHover.gif"});
var MenuBar4 = new Spry.Widget.MenuBar("MenuBar4", {imgDown:"SpryAssets/SpryMenuBarDownHover.gif", imgRight:"SpryAssets/SpryMenuBarRightHover.gif"});
//-->
</script>
</body>
</html>
```

3. 编写模型层代码

```java
package blog.bean.user;

public class Log {

    private String user;
    private String title;
    private String columns;
    private String content;
    private String permissions;
    private String comment;
    private String publishedTime, settop;
    public String getUser() {
        return user;
    }
    public void setUser(String user) {
        this.user = user;
    }
    public String getTitle() {
        return title;
    }
    public void setTitle(String title) {
        this.title = title;
    }
    public String getColumns() {
        return columns;
    }
    public void setColumns(String columns) {
        this.columns = columns;
    }
    public String getContent() {
        return content;
    }
    public void setContent(String content) {
        this.content = content;
    }
    public String getPermissions() {
        return permissions;
    }
    public void setPermissions(String permissions) {
        this.permissions = permissions;
    }
```

```java
        public String getComment() {
            return comment;
        }
        public void setComment(String comment) {
            this.comment = comment;
        }
        public String getPublishedTime() {
            return publishedTime;
        }
        public void setPublishedTime(String publishedTime) {
            this.publishedTime = publishedTime;
        }
        public String getSettop() {
            return settop;
        }
        public void setSettop(String settop) {
            this.settop = settop;
        }
    }
```

4. 编写控制层代码

```java
    package blog.servlet.user;

    import java.io.IOException;
    import java.io.PrintWriter;
    import java.util.List;

    import javax.servlet.ServletException;
    import javax.servlet.http.HttpServlet;
    import javax.servlet.http.HttpServletRequest;
    import javax.servlet.http.HttpServletResponse;

    import org.omg.CORBA.Request;

    import blog.Factory.DAOFactory;
    import blog.bean.user.Log;
    import blog.database.user.proxy.UserLogProxy;

    public class LogManagement extends HttpServlet {

        public void doGet(HttpServletRequest request, HttpServletResponse response)
            throws ServletException, IOException {
```

```java
        doPost(request, response);
    }

        @SuppressWarnings("unchecked")
        public void doPost(HttpServletRequest request, HttpServletResponse response)
                throws ServletException, IOException {

            String pageN = "1";
            int pageLimit = 10;
            int max = 1;
            int count;

            if(request.getParameter("pageN") != null) {
                pageN = request.getParameter("pageN");
            }

            String user = "a309131664";
              List<Log> logs = null;

              String columns = String.valueOf(1);

              UserLogProxy uLogProxy = DAOFactory.getUserLogProxy();
logs = uLogProxy.allfind(user, (Integer.parseInt(pageN) - 1) * pageLimit, pageLimit, columns);

                request.setAttribute("logs", logs);

request.getRequestDispatcher("user_log_management.jsp").forward(request, response);
        }

}

    package blog.servlet.user;

    import java.io.IOException;
    import java.util.List;

    import javax.servlet.ServletException;
    import javax.servlet.http.HttpServlet;
    import javax.servlet.http.HttpServletRequest;
    import javax.servlet.http.HttpServletResponse;

    import blog.Factory.DAOFactory;
    import blog.bean.user.Log;
```

```java
import blog.database.user.UserLog;
import blog.database.user.proxy.UserLogProxy;

public class NewLog extends HttpServlet {

//static int a = 88;

    public void doGet(HttpServletRequest request, HttpServletResponse response)
            throws ServletException, IOException {
    }

    public void doPost(HttpServletRequest request, HttpServletResponse response)
            throws ServletException, IOException {
        int Maxnum = 0;
        request.setCharacterEncoding("utf-8");
        UserLogProxy uLogProxy1 = DAOFactory.getUserLogProxy();
        List<Log> userLog = uLogProxy1.allfind("", 0, 100, "1");
        if((userLog.size() > 0) && (userLog! = null)) {
            Maxnum = Integer.parseInt(userLog.get(0).getUser());
        }
        String user = String.valueOf(Maxnum + 1);
        String title = request.getParameter("title");
        String columns = request.getParameter("columns");
        String content = request.getParameter("content");
        String permissions = request.getParameter("permissions");
        String comment = request.getParameter("comment");
        Log log = new Log();
        log.setColumns(columns);
        log.setComment(comment);
        log.setContent(content);
        log.setPermissions(permissions);
        log.setTitle(title);
        log.setUser(user);
        UserLogProxy uLogProxy = DAOFactory.getUserLogProxy();
        uLogProxy.insert(log);
        request.getRequestDispatcher("LogManagement")
            .forward(request, response);
    }

}


package blog.servlet.user;
```

```java
import java.io.IOException;
import java.io.PrintWriter;
import java.util.List;

import javax.servlet.ServletException;
import javax.servlet.http.HttpServlet;
import javax.servlet.http.HttpServletRequest;
import javax.servlet.http.HttpServletResponse;

import blog.Factory.DAOFactory;
import blog.bean.user.Log;
import blog.database.user.proxy.UserLogProxy;

public class EditorLog extends HttpServlet {

    public void doGet(HttpServletRequest request, HttpServletResponse response)
            throws ServletException, IOException {
        doPost(request, response);
    }

    public void doPost(HttpServletRequest request, HttpServletResponse response)
            throws ServletException, IOException {

        String user = request.getParameter("user");

        String url;
        if(request.getParameter("add") == null) {
            UserLogProxy uLogProxy = DAOFactory.getUserLogProxy();
            Log log = uLogProxy.logfind(user);
            request.setAttribute("log", log);
            request.setAttribute("user", user);
            UserLogProxy uLogProxy2 = DAOFactory.getUserLogProxy();
            url = "user_editorlog.jsp";
        } else {
            String title = new String(request.getParameter("title").getBytes("iso8859-1"), "utf-8");
            String columns = request.getParameter("columns");
            String content = new String(request.getParameter("content").getBytes("iso8859-1"), "utf-8");
            String permissions = request.getParameter("permissions");
            String comment = request.getParameter("comment");

            Log log = new Log();
```

```java
            log.setColumns(columns);
            log.setComment(comment);
            log.setContent(content);
            log.setPermissions(permissions);
            log.setTitle(title);
            log.setUser(user);
            UserLogProxy uLogProxy = DAOFactory.getUserLogProxy();
            uLogProxy.update(log);
            url = "LogManagement? user =" + user;
        }
        request.getRequestDispatcher(url).forward(request, response);
    }
}

package blog.servlet.user;

import java.io.IOException;
import java.io.PrintWriter;

import javax.servlet.ServletException;
import javax.servlet.http.HttpServlet;
import javax.servlet.http.HttpServletRequest;
import javax.servlet.http.HttpServletResponse;

import blog.Factory.DAOFactory;
import blog.bean.user.Log;
import blog.database.user.proxy.UserLogProxy;

public class DeleteLog extends HttpServlet {
    public void doGet(HttpServletRequest request, HttpServletResponse response)
            throws ServletException, IOException {
        doPost(request, response);
    }
    public void doPost(HttpServletRequest request, HttpServletResponse response)
            throws ServletException, IOException {
        String user = request.getParameter("user");

        UserLogProxy uLogProxy = DAOFactory.getUserLogProxy();

        uLogProxy.delete(user);

        request.getRequestDispatcher("LogManagement").forward(request, response);
    }
}
```

}

```
package blog.inter.user;

import java.util.List;

import blog.bean.user.Log;

public interface ILog {
    //插入
    public void insert(Log log);
    //删除
    public void delete(String user);
    //更新
    public void update(Log log);
    //查看
    public Log logfind(String user);
    //查看所用 日志
    public List<Log> allfind(String user, int pageN, int pageLimit, String columns);
}
```

任务十二：物流管理系统的公司信息管理

一、任务描述

你作为《物流管理系统》项目开发组的程序员，请实现如下功能：
➢ 显示公司信息列表；
➢ 添加公司信息；
➢ 查询公司信息。

二、功能描述

（1）点击图12.1所示页面顶部导航条中的"承运管理"菜单项，在打开的左侧菜单中点击"公司信息"菜单项，则在右边的主体部分显示公司信息列表。

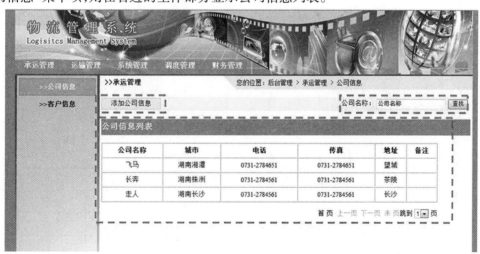

图12.1 公司信息列表页面

（2）在图12.1中，输入需查询的公司名称，点击"查找"按钮，将查询结果显示在公司列表中。

（3）在图12.1中，点击"添加公司信息"按钮，则跳转到公司信息录入页面，如图12.2所示。

图 12.2　公司信息录入页面

（4）点击图 12.2 中的"保存"按钮，对图中打"＊"号的输入部分进行必填校验，通过校验后在数据库中添加公司信息。

（5）公司信息增加成功后，跳转到图 12.1 所示页面，显示更新后的公司信息列表。

（6）测试程序，在添加公司信息页面新增两条以上公司信息。

三、要求

1. 界面实现

以提供的素材为基础，实现图 12.1、图 12.2 所示页面。

2. 数据库实现

（1）创建数据库 LogisticsDB。

（2）创建公司信息表（T_logistics_company），表结构见表 12.1。

表 12.1　公司信息表（T_logistics_company）表结构

字段名	字段说明	字段类型	允许为空	备注
Company_id	公司编号	varchar(16)	否	主键
Company_name	公司名称	varchar(60)	否	
Company_city	所在城市	varchar(20)	是	
Company_phone	联系电话	varchar(20)	是	
Company_fax	传真	varchar(20)	是	
Company_adress	地址	varchar(100)	是	
Company_remark	备注	varchar(500)	是	

（3）在表 T_logistics_company 插入以下记录，见表 12.2。

表 12.2　公司信息表(T_logistics_company)记录

Company_id	Company_name	Company_city	Company_phone	Company_fax	Company_adress	Company_remark
2011-01	飞马	湖南湘潭	0731-52584651	0731-52584651	湘乡	
2011-02	长奔	湖南株洲	0731-23553378	0731-23553378	茶陵	
2011-03	旭日	湖南长沙	0731-82788879	0731-82788879	长沙	

3. 功能实现

(1)功能需求如图 12.3 所示。

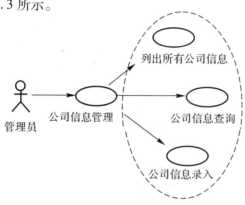

图 12.3　公司信息模块用例图

(2)依据公司信息列表活动图完成公司信息列表显示功能,如图 12.4 所示。

(6)依据添加公司信息活动图完成添加公司信息功能,如图 12.5 所示。

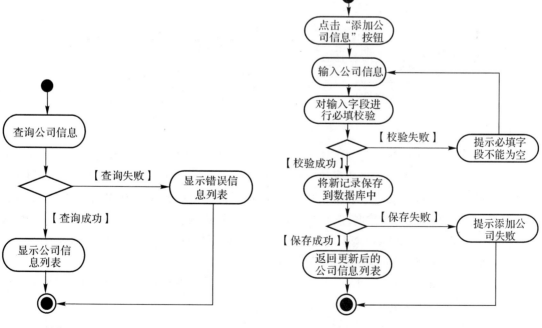

图 12.4　公司信息列表活动图　　　　图 12.5　添加公司信息活动图

四、必备知识

1. HTML ＜iframe＞ 标签

IFRAME,HTML 标签,作用是文档中的文档,或者浮动的框架(FRAME)。所有浏览器都支持 ＜iframe＞ 标签,iframe 元素会创建包含另外一个文档的内联框架(即行内框架)。

(1)定义和用法。iframe 元素会创建包含另外一个文档的内联框架(即行内框架)。

(2)核心属性见表 12.3。

表 12.3　核心属性

属性	值	描述
class	classname	规定元素的类名(classname)
id	id	规定元素的唯一 id
style	style_definition	规定元素的行内样式(inline style)
title	text	规定元素的额外信息(可在工具提示中显示)

(3)语言属性见表 12.4。

表 12.4　语言属性

属性	值	描述
dir	ltr | rtl	设置元素中内容的文本方向。
lang	language_code	设置元素中内容的语言代码。
xml:lang	language_code	设置 XHTML 文档中元素内容的语言代码。

(4)键盘属性见表 12.5。

表 12.5　键盘属性

属性	值	描述
accesskey	character	设置访问元素的键盘快捷键。
tabindex	number	设置元素的 Tab 键控制次序。

2. 三层结构概述

在操作之前,根据答题要求,需要用三层结构来实现所有功能。三层结构是一种严格分层方法,即数据访问层(DAL)只能被业务逻辑层(BLL)访问,业务逻辑层只能被表示层(USL)访问,用户通过表示层将请求传送给业务逻辑层,业务逻辑层完成相关业务规则和逻辑,并通过数据访问层访问数据库获得数据,然后按照相反的顺序依次返回将数据显示在表示层。有的三层结构还加了 Factory,Model 等其他层,实际都是在这三层基础上的一种扩展和应用。

三层结构(从上到下):

(1)表示层(USL):主要表示 WEB 方式,也可以表示成 WINFORM 方式。如果逻辑层相当强大和完善,无论表现层如何定义和更改,逻辑层都能完善地提供服务。

(2)业务逻辑层(BLL):主要是针对具体的问题的操作,也可以理解成对数据层的操作,对数据业务逻辑处理。如果说数据层是积木,那逻辑层就是对这些积木的搭建。

(3)数据访问层(DAL):主要是对原始数据(数据库或者文本文件等存放数据的形式)的操作层,具体为业务逻辑层或表示层提供数据服务.

在本题和以后的所有题目中,数据库访问层都是大同小异,都是把对数据库操作的共同部分提取出来,封装到一个类中,以后就可以调用类中的方法,轻松的实现的操作,并把编程的精力集中在实现在应用逻辑和页面实现上.

数据访问层的添加:在解决方案资源管理器中,右键单击方案的名字,并在弹出的窗口中选中"添加新项",如图12.6所示。

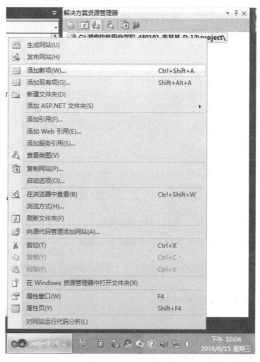

图12.6　填加新项

在弹出窗口的中间选中"类",将该类命名为"DbHelper.cs",如图12.7所示。

图12.7　选中"类"

在弹出的窗口中单击"是",如图 12.8 所示。

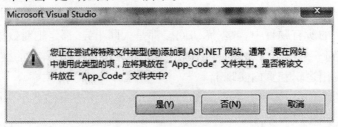

图 12.8 单击"是"

类 DbHelper.cs 的内容如下:
```
using System.Data.SqlClient;
public class DBHelper
{
    public string str = "Data Source = localhost; Initial Catalog = LogisticsDB ; Integrated Security = True";
    //执行 SQL 语句返回受影响的行数
    public int Operate(string sql)
    {
        using (SqlConnection con = new SqlConnection(str))
        using (SqlCommand com = new SqlCommand(sql, con))
        {
            con.Open();
            return com.ExecuteNonQuery();
        }
    }
    //执行 SQL 返回结果集的首行首列
    public object GetScalar(string sql)
    {
        using (SqlConnection con = new SqlConnection(str))
        using (SqlCommand com = new SqlCommand(sql, con))
        {
            con.Open();
            return com.ExecuteScalar();
        }
    }
    //执行 SQL 语句返回一个数据读取器(SqlDataReader)
    public SqlDataReader GetReader(string sql)
    {
        using (SqlConnection con = new SqlConnection(str))
        using (SqlCommand com = new SqlCommand(sql, con))
        {
            con.Open();
```

```
            return com.ExecuteReader(CommandBehavior.CloseConnection);
        }
    }
    //执行SQL语句返回一个数据集(DataSet)
    public DataSet FillDataSet(string sql)
    {
        using (SqlConnection con = new SqlConnection(str))
        {
            SqlDataAdapter sda = new SqlDataAdapter(sql, con);
            DataSet ds = new DataSet();
            sda.Fill(ds);
            return ds;
        }
    }
}
```

创建好数据访问层以后,接下来是业务逻辑层,业务逻辑层主要负责对数据层的操作,本题中,业务逻辑主要是建设工程项目信息的列表显示和建设工程项目信息的添加。首先添加业务逻辑层,操作步骤如下:

在解决方案资源管理器中,右键单击方案的名字,并在弹出的窗口中选中"添加新项",并在弹出的"添加新项"窗口中选中"类",可以将类名更改为companyBLL.CS或者跟BLL相关的名字,如图12.9所示。

图12.9　更改类名

业务逻辑层在体系架构中的位置很关键,它处于数据访问层与表示层中间,起到了数据交换中承上启下的作用。本项目中,业务逻辑主要实现建设工程项目信息的列表显示和建设工程项目信息的添加。在添加实际功能之前,因为本类需要使用数据库类中的对象,所以应该在最上面添加一句话。

```csharp
using System.Data.SqlClient;
```
然后根据具体的功能,分别添加代码如下:
```csharp
public class CompanyBLL
{
    //查询全部或者部分公司信息列表
    public DataSet getcompany(string key)
    {
        string sql = string.Format("select * from T_logistics_company where Company_name like '%{0}%'", key);
        DBHelper db = new DBHelper();
        return db.FillDataSet(sql);
    }

    //添加公司信息
    public int addcompany(string Company_id, string Company_name, string Company_city, string Company_phone, string Company_fax, string Company_address, string Company_remark)
    {
        string sql = string.Format("insert into T_logistics_company(Company_id,Company_name,Company_city,Company_phone,Company_fax,Company_address,Company_remark) values('{0}','{1}','{2}','{3}','{4}','{5}','{6}'", Company_id, Company_name, Company_city, Company_phone, Company_fax, Company_address, Company_remark);
        DBHelper db = new DBHelper();
        return db.Operate(sql);
    }
}
```

五、解题思路

根据技能抽查的答题要求,答案以"答题文件"的形式提交。请按以下要求创建答题文件夹和答题文件:

1. 创建答题文件夹

创建以"所属学校名_身份证号_姓名_题号"命名的文件夹,存放所有答题文件,例如:"湖南软件职业学院_430101***********_李维_A_1\"。

2. 创建答题文件

(1)项目源文件。创建 project 子文件夹,如:"湖南软件职业学院_430101*******_李维_A_1\project\",存放项目所有源代码。

(2)数据库备份文件。创建 bak 子文件夹,如:"湖南软件职业学院_430101*******_李维_A_1\bak\",存放数据库备份文件,它用于教师阅卷时,还原运行环境。无数据库备份文件,则扣除相应的技术分。

(3)页面截图文件。创建 picture 子文件夹,如:"湖南软件职业学院_430101*******_李维_A_1\picture\",存放截图.doc 文件,它用于保存程序运行过程中的屏幕截图,每张截图必须有文字说明,要求每个实现的功能至少截两张图,如"新增工程"功能,要求有"新增"之前的截图和"新增"成功后的截图。

六、操作步骤

根据系统提供的静态页面,可以看出,整个任务是用框架技术来实现的。表12.6是系统提供的静态页面的名字以及功能说明。

表12.6 静态页面名字与功能

序号	素材名	功能说明	备注
1	Index.html	物流管理系统首页	开始页面
2	Main_client01.html	公司信息列表页面	
3	Mian_client02.html	添加公司信息页面	
4	Left_Continue.html	左边导航页面	
5	Left_notPage	左边无导航信息页面	
6	css	页面中样式文件夹	
7	images	页面中图片文件夹	

我们只需要将Main_client01.html和Mian_client02.html两个页面做成aspx页面,再嵌到对应的静态页面中去就可以了。

(1)创建数据库LogisticsDB和数据表T_logistics_company

抽查时会提供数据库文件,只需要将文件附加到SQL服务器中,或者将提供的BAK还原到当前SQL服务器中即可。

第二步,编写代码实现公司信息列表显示

(1)首先打开VS,新建项目,将项目位置定位于答题文件夹project,并将所有的静态页面复制到解决方案资料管理器中,如图12.10所示。

图12.10 静态页面的复制

(2) 新建 Web 窗体,命名为 project_list.aspx。如图 12.11 所示。

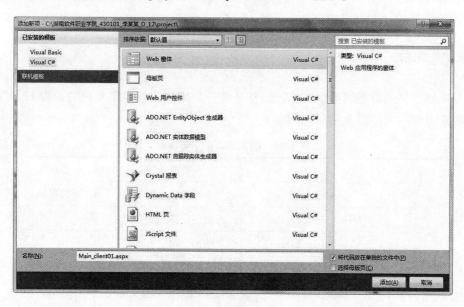

图 12.11

然后将 Main_client01.html 页面源文件中的 < head > </head > 和 < body > </body > 这 2 组标签的全部内容复制到 Main_client01.aspx 页面中对应的位置,这样充分利用静态页面的布局和内容。

这样操作之后,Main_client01.aspx 页面的变成图 12.12 所示:

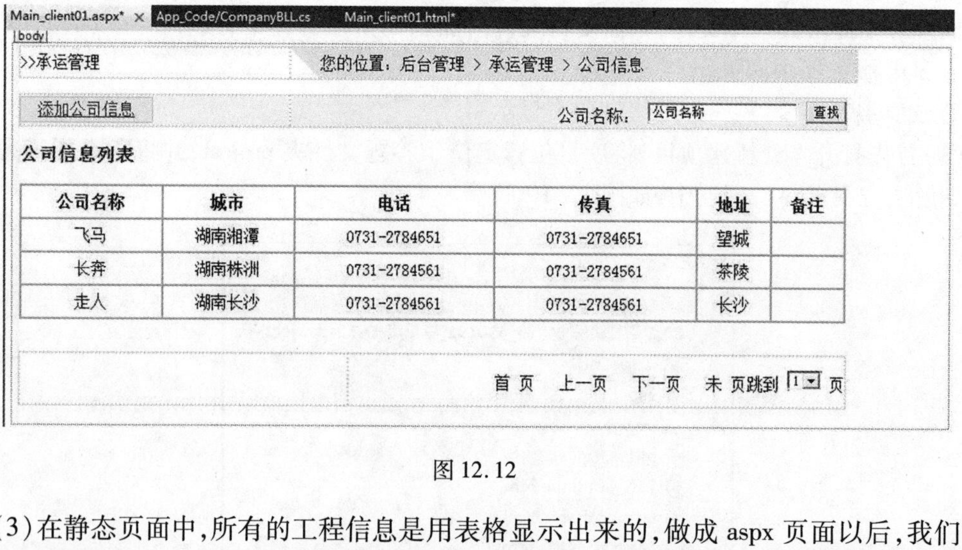

图 12.12

(3) 在静态页面中,所有的工程信息是用表格显示出来的,做成 aspx 页面以后,我们需要用数据库控件来显示。首先我们将工具箱中的 gridview 控件拖到页面中的静态表格的下面,如图 12.13 所示。

然后通过后台代码,将数据库中的所有公司信息显示在数据源控件上。这段代码应该写在 Page_Load 事件中。

— 258 —

任务十二：物流管理系统的公司信息管理

图 12.13

```
protected void Page_Load(object sender, EventArgs e)
{//页面加载的时候,将所有的公司信息呈现在页面控件上
    CompanyBLL tc = new CompanyBLL();
    GridView1.DataSource = tc.getcompany("");
    GridView1.DataBind();
}
```

运行之后,就是图 12.14 的这种效果。

图 12.14

大家可能还是会发现，从数据库中查询出来的表格显示的列名变是数据库的字段名，并且我们也不需要那么多列，现在我们来进行修改。点击控件右上角的小三角按钮，然后点击"编辑列"，如图 12.15 所示。

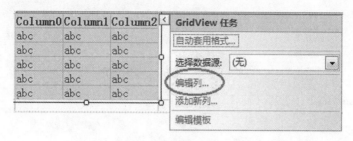

图 12.15

在弹出窗口中分别添加 6 个 BoundFiled 控件，并去掉窗口左下角的"自动生成字段"即不勾选；如图 12.16 所示。

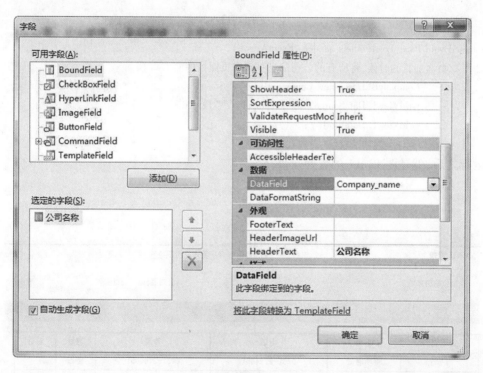

图 12.16

注意，具体设置是——选中 BoundFiled 点击"添加"，然后在右边填写数据中的 DataField（数据源中的列名，例如当前的编号 Company_name，外观中的 HeaderText（要显示在页面上的列名，例如当前的"公司名称"），然后取消"自动生成字段"的勾选框，最后点击"确定"，相关的页面效果变成图 12.17 所示。

点击"自动套用格式"可以根据需要设置相应的样式，点击"编辑模板"进行操作中的模板设置，如图 12.18 所示。

图 12.17

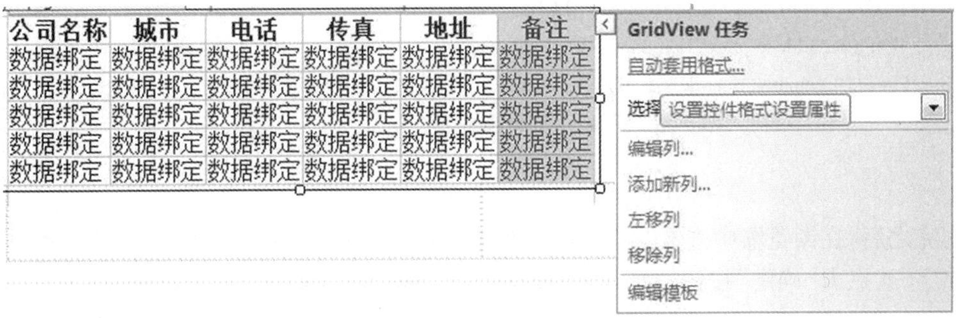

图 12.18

再经过一些细微的调整,将静态的表格数据删除,调整 gridview 各列的宽度,并设置文本居中对齐,即可以得如图 12.19 的界面。

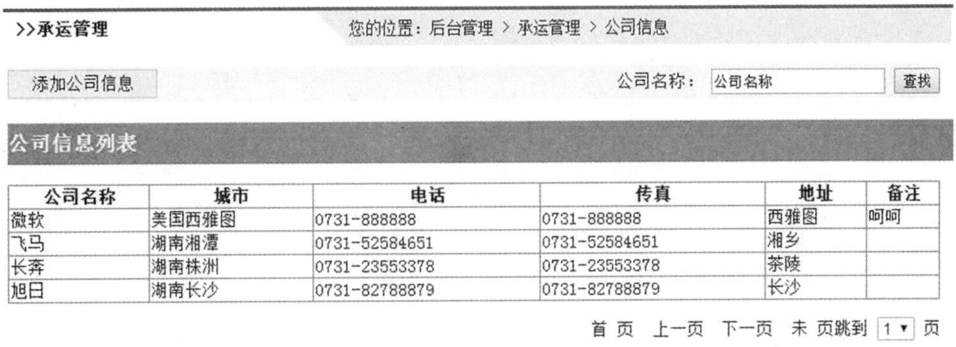

图 12.19

第三步,编码实现公司信息的添加

本功能的实现如同上例,需要添加同名的页面 Mian_client02.aspx,,需要借用静态页面提供的 HTML 源码,然后在本页面中,将所有的 7 个 TextBox 和 1 个 Button 替换成服务端的控件，并且将页面中的"＊"换成验证控件 RequiredFieldValidator，与待验证的文本框一一对应。

然后，在"确定"按钮的点击事件中，写好以下代码完成公司信息的添加

```
protected void Button1_Click(object sender, EventArgs e)
{//添加公司信息
    if(Page.IsValid)//如果页面验证通过
    {
        string Company_id, Company_name, Company_city, Company_phone, Company_fax, Company_address, Company_remark;
        Company_id = TextBox1.Text;
        Company_name = TextBox2.Text;
        Company_city = TextBox3.Text;
        Company_phone = TextBox4.Text;
        Company_fax = TextBox5.Text;
        Company_address = TextBox6.Text;
        Company_remark = TextBox7.Text;
        CompanyBLL tc = new CompanyBLL();
        if(tc.addcompany(Company_id, Company_name, Company_city, Company_phone, Company_fax, Company_address, Company_remark) > 0)
            Response.Redirect("Main_client01.aspx");
    }
}
```

写完之后，在浏览器中查看"Mian_client02.aspx"页面。如果用户没有输入公司编号或者公司名字，就点击"确定"按钮，对应的验证控件就会弹出错误提示"＊"；如图12.20所示。

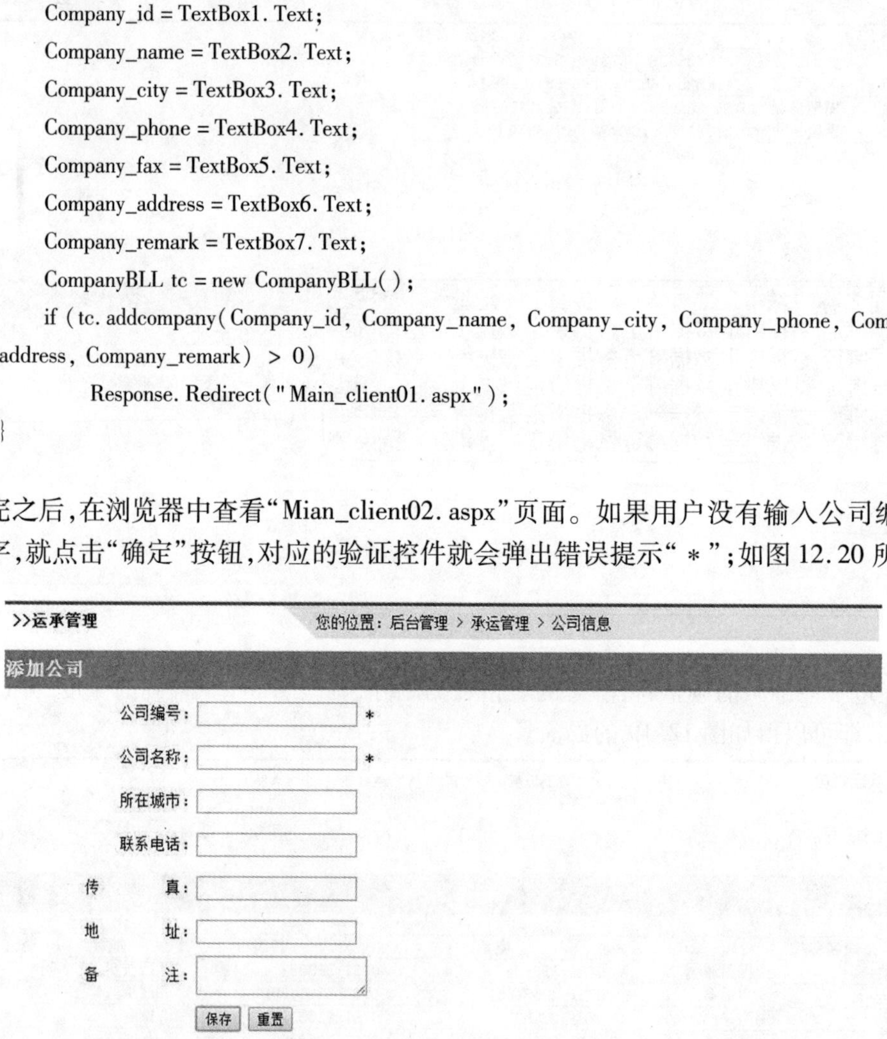

图 12.20

如果将所有的信息都输入正确，则会自动跳转到 project_list.aspx 页面，并显示更新后的项目工程列表，如图 12.21 所示。

第四步，编码实现查询客户信息

在图12.21中，输入需查询的公司名称，点击"查找"按钮，将查询结果显示在公司列表，该操作跟前面的操作非常类似，将文本框和换成替换成 asp 服务器端控件，并为"查找"的单击

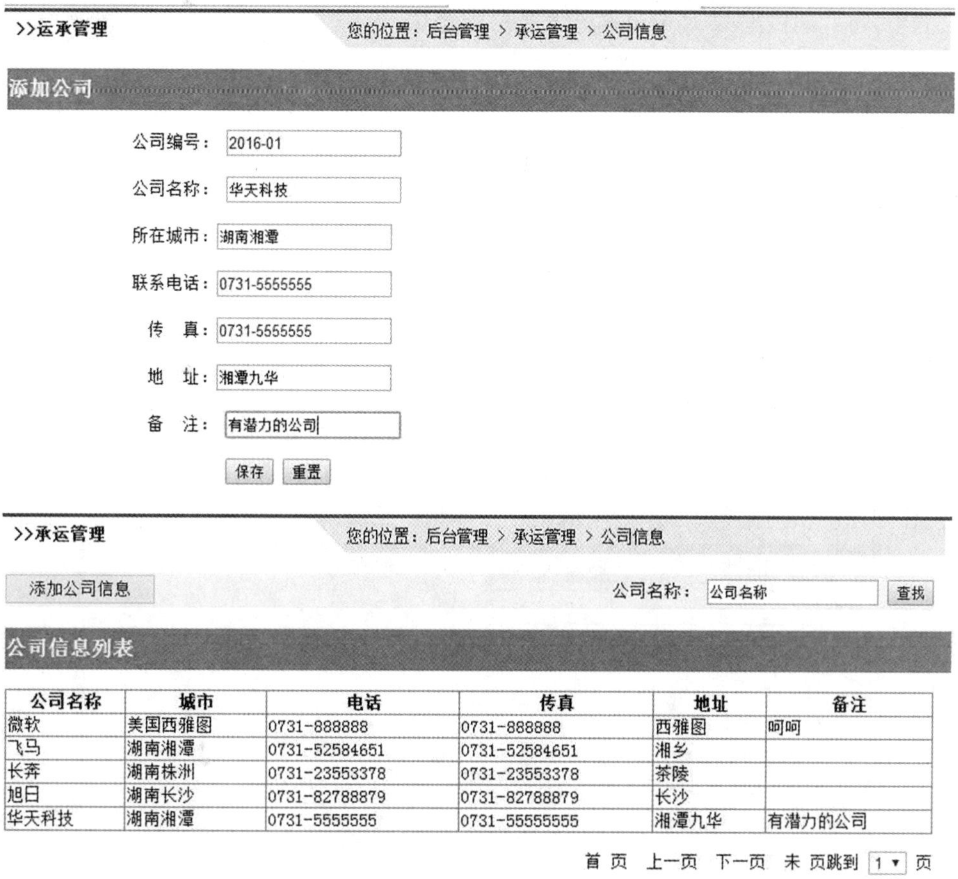

图 12.21

事件撰写代码如下：

```
protected void Button1_Click(object sender, EventArgs e)
    {//按关键字查找公司信息
        CompanyBLL cb = new CompanyBLL();
        GridView1.DataSource = cb.getcompany(TextBox1.Text.Trim());
        GridView1.DataBind();
    }
```

然后,可以在浏览器中看到如图 12.22 所示的效果。

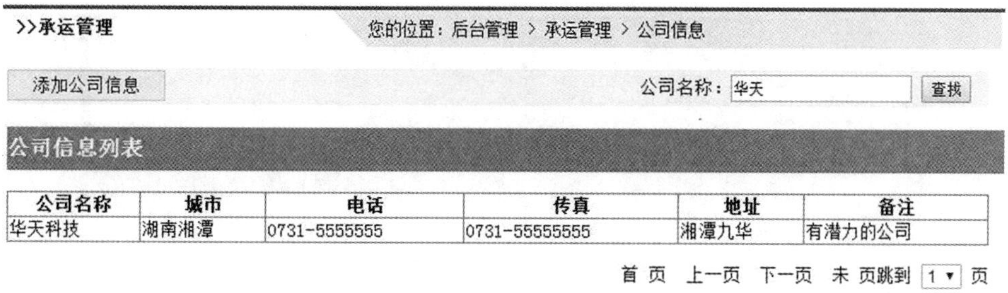

图 12.22

第五步,将我们做好的 aspx 页面嵌入到框架当中

通过对素材的浏览,我们知道,用户点击 index.html 左边左侧菜单中点击"公司信息"菜单项才可以看到所有公司信息列表,则找到对应的子页面 Left_Continue.html,将该链接的目标地址改成 Main_client01.aspx,即

<li> <a href = "Main_client01.aspx" target = "main" >&gt;&gt 公司信息 </a> </li>

用户在该页面通过点击"添加公司信息"的按钮跳转到 Main_client02.aspx,则可以通过源代码将"添加公司信息"按钮的目标地址定位到 Main_client02.aspx,即

<a href = "Main_client02.aspx" target = "main" > 添加公司信息 </a> </div> </td>

如此,整个题目的功能完成。

第六步,功能测试与网站发布

确定好网站的所有功能和逻辑都实现了以后,可以进行网站的生成和发布。

(1)生成网站。在待发布的网站的解决方案资源管理器上,选中整个解决方案,并单击鼠标右键,在弹出的快捷菜单中选中"生成网站"。如图 12.23 所示。

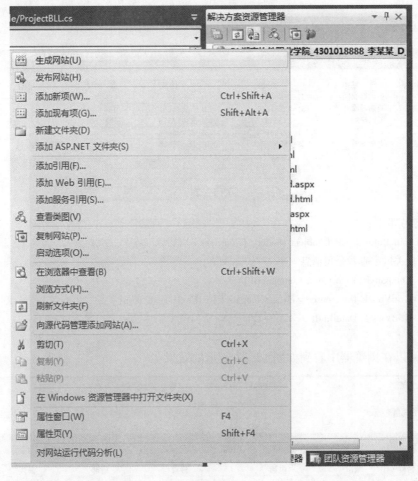

图 12.23

等到状态栏的左下方提示"生成成功",再选中整个解决方案,并单击鼠标右键,在弹出的快捷菜单中选中"发布网站"。如图 12.24 所示。

任务十二:物流管理系统的公司信息管理

图 12.24

(2)发布。在弹出的"发布网站"的窗口中,输入目标位置,比如我这里是"c:\MyWeb",然后"确定"。等到状态栏的左下方提示"发布成功",就可以进行下一步的操作了。

下面就把发布出来的网站挂到 IIS 上,控制面板→管理工具→Internet 信息服务(IIS)管理器,如图 12.25 所示。

图 12.25

在树列表中选择网站→右键→添加网站,如图12.26所示。

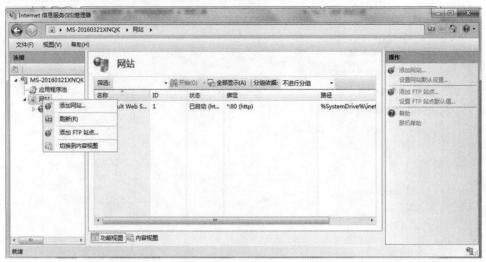

图12.26

填上相应的网站名称(自定义)、选择应用程序池、物理路径(刚才发布的网站的路径),选上IP地址(IPv4),如图12.27所示。

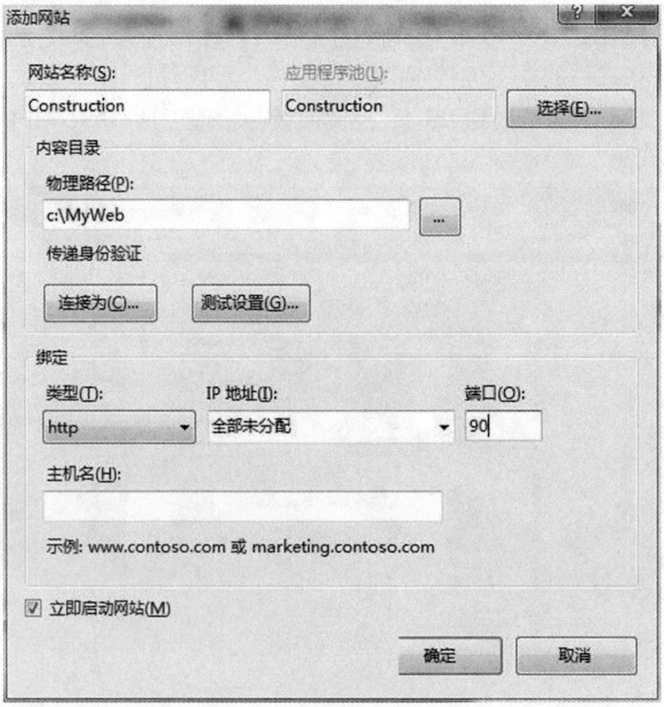

图12.27

若端口号已经绑上了其他网站,系统会提示是否绑在同一端口,最好再选一个端口。然后切换到,右键浏览网站,这样就OK了。

发布中的常见问题:

（1）配置错误，如图12.28所示。

图12.28

分析：主要原因IIS中是应用程序池的版本与你开发使用的.netFramework版本不一致。一般一个网站对应一个应用程序池，并与网站同名。visual studio 2010默认安装对应的.netFramework是4.0，而本机上一般是2.0版本。所以不匹配。

解决方法：在IIS中点击应用程序池，找到你的网站的应用程序池，右键高级设置.netFramework改成V4.0就可以了。

再次打开这个网站，就没有这样的错误了。再次打开，也可以会有下面的错误出现。

（2）HTTP错误，如图12.29所示。

图12.29

错误分析：vs2010默认采用的是.NET 4.0框架，4.0框架是独立的CLR，和.NET 2.0的不同，如果想运行.NET 4.0 框架的网站，需要用 aspnet_regiis 注册.NET 4.0 框架，然后用.NET 4.0 框架的 class 池，就可以运行.NET 4.0 框架的 web 项目了。

造成上述错误的原因极有可能是：由于先安装.NetFramework v4.0 后安装 iis 7.5 所致。

如何用 aspnet_regiis 注册 4.0 框架？其方法如下：

1）找到.NET 4.0 框架下 aspnet_regiis 所在目录，在 C 盘根目录中搜索 aspnet_regiis，找到 4.0 框架下 aspnet_regiis 的目录位置，如图 12.30 所示。

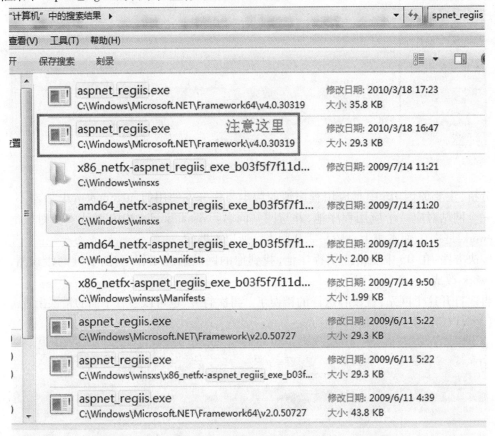

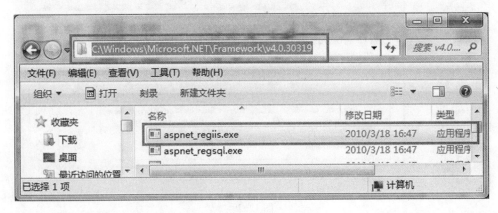

图 12.30

2)以管理员的身份运行 DOS 命令行,执行"开始→所有程序→附件→命令提示符(右击选择'以管理员身份运行(A)')",弹出"管理员:命令提示符"窗口,如图 12.31 所示。

图 12.31

3)执行命令"cd C:\Windows\Microsoft.NET\Framework\v4.0.30319",进入到"C:\Windows\Microsoft.NET\Framework\v4.0.30319"目录,如图 12.32 所示。

图 12.32

然后执行命令"aspnet_regiis.exe -i",注册"aspnet_regiis",稍等片刻,aspnet_regiis 成功注册如图 12.33 所示。

图 12.33

看到图中的界面,就可以在 IIS 中运行.net4.0 部署的网站喽!

任务十三：物流管理系统的客户信息管理

一、任务描述

你作为《物流管理系统》项目开发组的程序员，请实现如下功能：
- 显示客户信息列表；
- 查询客户信息；
- 修改客户信息。

二、功能描述

(1)点击图 13.1 所示页面顶部导航条中的"承运管理"菜单项，在打开的左侧菜单中点击"客户信息"菜单项，则在右边的主体部分显示客户信息列表。

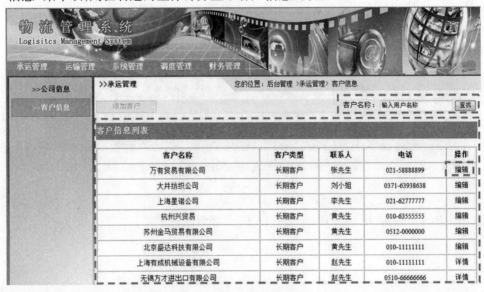

图 13.1　客户信息列表页面

(2)在图 13.1 中，点击操作列中的"编辑"按钮，跳转到客户信息修改页面，修改所在行的客户信息，如图 13.2 所示。

图 13.2 客户信息修改页面

(3)点击图 13.2 中的"保存"按钮,对图中打"﹡"号的输入部分进行必填校验,通过校验后,数据库中修改客户信息。

(4)修改客户信息成功后,跳转到图 13.1 所示页面,显示更新后的客户信息列表。

三、要 求

1. 界面实现

以提供的素材为基础,实现图 13.1、图 13.2 所示页面。

2. 数据库实现

(1)创建数据库 LogisticsDB。

(2)创建客户信息表(T_logistics_client_info),表结构见表 13.1。

表 13.1 客户信息表(T_logistics_client_info)表结构

字段名	字段说明	字段类型	允许为空	备注
Client_info_id	客户编号	varchar(10)	否	主键
Client_info_name	客户名称	varchar(50)	否	
Client_info_type	客户类型	varchar(20)	否	
Client_info_contacts	联系人	varchar(20)	否	
Client_info_phone	联系电话	varchar(20)	否	
Client_info_address	联系地址	varchar(100)	是	
Client_info_remark	备注	varchar(100)	是	

(3)在表 T_logistics_client_info 插入以下记录,见表 13.2。

表13.2 客户信息表(T_logistics_client_info)记录

Client_info_id	Client_info_name	Client_info_type	Client_info_contacts	Client_info_phone	Client_info_adress	Client_info_remark
CI0001	万有贸易有限公司	长期客户	张先生	021-28888899	上海	
CI0002	大井纺织公司	长期客户	刘小姐	0731-53938638	望城	
CI0003	上海星诺公司	长期客户	李先生	021-62777777	上海	

3. 功能实现

(1)功能需求如图13.3所示。

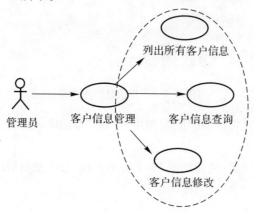

图13.3 客户信息管理模块用例图

(2)依据公司信息列表活动图完成客户信息列表显示功能,如图13.4所示。

(3)依据修改客户信息活动图完成修改客户信息功能,如图13.5所示。

(4)修改客户信息页面中客户类型下拉框的值为{"长期客户","短期客户"}。

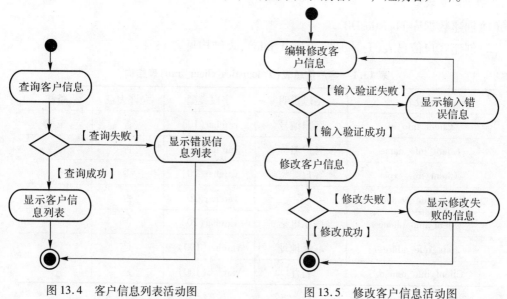

图13.4 客户信息列表活动图　　图13.5 修改客户信息活动图

四、必备知识

1. 数据库相关知识

（1）使用 MS SQL Server 2005/2008 创建数据库，创建数据表，设置表的字段，数据类型，主键，外键，约束。

（2）向数据表插入、删除、修改、查询数据。

2. 页面相关知识

（1）看懂系统所提供的素材，并理解框架技术。

（2）使用 asp.net 验证控件对页面必要的内容进行校验。

3. Asp.net 相关知识

（1）理解如何将 aspx 页面嵌入到框架中。

（2）使用数据源绑定控件显示数据库的数据。

（3）合理的使用转发和重定向控制项目的页面跳转。

（4）使用 ado.net 技术与数据库进行交互。

（5）Asp.net 三层结构的数据访问层、业务逻辑层和表示层的功能和合作。

（6）GridView 控件中自定义列的增加。

GridView 可以根据数据源自动生成列，但是如果我们需要自定义列的显示方式，让 GridView 的列完完全全的由我们自己来控制，我们就需要用到一种特殊的列——TemplateField。因为 GridView 生成的列都是一个字段一列，如果我们需要把两个字段合并为一列显示呢？我们可以使用模板列，指定包含标记和控件的模版，自定义列的布局和行为。

自定义操作步骤如下：

点击 GridView 控件的右上角的控件按钮，在弹出的快捷菜单中点击"编辑列"，如图 13.6、图 13.7 所示。

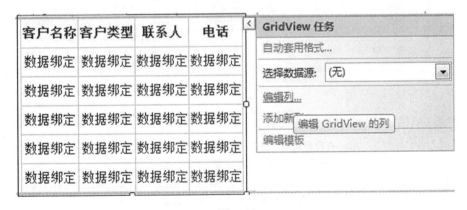

图 13.6

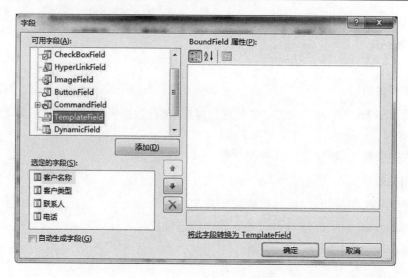

图 13.7

选中 TemplateField,并点击"添加",然后在右边的属性框中设置 HeaderText 的值为"操作",如图 13.8 所示。

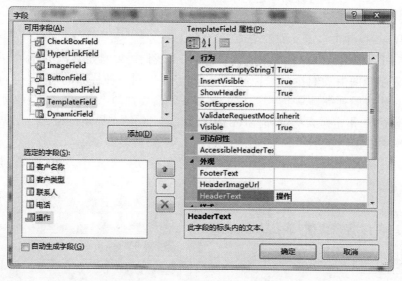

图 13.8

点击"确定"按钮以后,GridView 控件变成如图 13.9 所示。

客户名称	客户类型	联系人	电话	操作
数据绑定	数据绑定	数据绑定	数据绑定	
数据绑定	数据绑定	数据绑定	数据绑定	
数据绑定	数据绑定	数据绑定	数据绑定	
数据绑定	数据绑定	数据绑定	数据绑定	
数据绑定	数据绑定	数据绑定	数据绑定	

图 13.9

然后再点击 GridView 控件的右上角的控制按钮,在弹出的快捷菜单中选中"编辑模板",如图 13.10 所示。

图 13.10

在工具栏上拖一个 LinkButton 放到 ItemTemplate 中,并将 Text 属性改成"编辑","结束模板编辑"以后就 OK。如图 13.11 所示。

(7) LinkButton 的 CommandArgument 属性。你也许对 LinkButton 组件中 CommandArgument 属性有点迷惑,他们到底会有什么作用。其实他们有着非常重要的作用。举例如下:

假如在你的一个页面中有 100 个记录,就会有 100 个相同的"编辑"链接按钮,那么在服务器端要判断到底是哪个 LinkButton 被按动了。如果没有这个属性,可以通过给每一个 LinkButton 的 Text 属性赋值,然后通过检测 Text 属性值,来判断到底是哪个 LinkButton 被按动了。当然你也可以通过给每一个 LinkButton 命不同的名(当然你就是想命同样的名,在 ASP.NET 中也是不可以的),然后在事件处理函数中,通过检测 Sender 属性值来判断哪个 LinkButton 被按动。虽然上述这些方法都是可行的,但无一不繁琐。

有了 CommandArgument 属性就可以方便的解决这些问题。在程序中给 LinkButton 的 CommandArgument 属性赋值,然后通过 OnCommand 事件中,就可以得到从 CommandEventArgs 类中的数据,从而到判断是哪个 LinkButton 被按动了。那么 CommandEventArgs 类有是干什么的? CommandEventArgs 类存储了和按钮(Button)事件相关的数据,并且可以在事件处理中通过 CommandEventArgs 类的属性来访问这些数据。说的明白些,就是当 LinkButton 被按动后,这个 LinkButton 所触发的数据都被储存到服务器的 CommandEventArgs 类中,访问 CommandEventArgs 类中的属性也就访问了被按动的 LinkButton 了。

比如本例中如图 13.12 所示。

客户名称	客户类型	联系人	电话	操作
万有贸易有限公司	长期客户	张先生	021-2888899	编辑
大井纺织公司	长期客户	刘小组	0731-53938638	编辑
aa	bb	cc	dd	编辑

图 13.12

有多少条记录就会有多少个"编辑"链接按钮,如果知道用户是点击哪一行的"编辑"按钮呢,就需要 CommandArgument 属性。要将当前记录的主键值作为参数赋给"编辑"按钮,具体做法如下:

切换到源代码,当前 GridView 控件的源代码如下:

```
<asp:GridView ID="GridView1" runat="server" AutoGenerateColumns="False">
    <Columns>
        <asp:BoundField DataField="Client_info_name" HeaderText="客户名称" />
        <asp:BoundField DataField="Client_info_type" HeaderText="客户类型" />
        <asp:BoundField DataField="Client_info_contacts" HeaderText="联系人" />
        <asp:BoundField DataField="Client_info_phone" HeaderText="电话" />
        <asp:TemplateField HeaderText="操作">
            <ItemTemplate>
                <asp:LinkButton ID="LinkButton1" runat="server">编辑</asp:LinkButton>
            </ItemTemplate>
        </asp:TemplateField>
    </Columns>
</asp:GridView>
```

在"编辑"控件的那一行,加一个属性,将当前记录的主键值设置为当前 LinkButton 的事件参数,修改如下:

```
<asp:GridView ID="GridView1" runat="server" AutoGenerateColumns="False">
    <Columns>
        <asp:BoundField DataField="Client_info_name" HeaderText="客户名称" />
        <asp:BoundField DataField="Client_info_type" HeaderText="客户类型" />
        <asp:BoundField DataField="Client_info_contacts" HeaderText="联系人" />
        <asp:BoundField DataField="Client_info_phone" HeaderText="电话" />
        <asp:TemplateField HeaderText="操作">
            <ItemTemplate>
                <asp:LinkButton ID="LinkButton1" runat="server" CommandArgument='<%#Eval("Client_info_id") %>'>编辑</asp:LinkButton>
            </ItemTemplate>
        </asp:TemplateField>
    </Columns>
</asp:GridView>
```

接下来的操作中,点击"编辑"按钮产生的事件不是 OnClick,而应该是 OnCommand。因为 Command 事件的处理程序可以带一个 CommandEventArgs 类型的参数,该参数包含被点击的按钮的 CommandArgument 属性的值,这使你可以在一个事件处理程序中区分不同的引发事件的对象和参数,而 Click 事件处理程序则接收一个 EventArgs 类型的参数,该参数是空值。

五、解题思路

1. 数据库思路

(1)根据项目要求创建数据库和数据表,向数据表中插入合适的测试数据。
(2)添加数据访问层,并在其中写好连接字符串和若干个通用的数据库操作方法
(3)添加业务逻辑层,设置适当的参数,实现本任务所要求的全部功能,本层可以引用数据访问层的方法。

2. 表示层思路

(1)将提供的素材页面改写为 aspx 页面。
(2)将数据库数据的显示控件和部分表单控件改成 asp 服务器端控件。

3. 实现思路

（1）在表示层访问业务逻辑层，按照题目要求调用业务逻辑层的方法。

（2）数据添加的时候，正确获取用户在表单中提取的数据，并写入数据库中，然后进行页面跳转逻辑控制。

（3）查询的时候，调用业务逻辑层的方法，获取数据库中的数据，并按照用户要求的格式显示出来。

六、操作步骤

业务逻辑层代码：

```
public DataSet getclient(string key)
    {
        string sql = string.Format("select * fromT_logistics_client_info where Client_info_name like '%{0}%'",key);
        DBHelper db = new DBHelper();
        return db.FillDataSet(sql);
    }
//更新客户信息
    public intupdateclient (string cid, string cname, string ctype, string contacts, string phone, string address, string remark)
    {
        string sql = string.Format(update T_logistics_client_info set Client_info_name='{0}',Client_info_type='{1}',Client_info_contacts='{2}',Client_info_phone='{3}',Client_info_address='{4}',Client_info_remark='{5}' where Client_info_id='{6}'", cname, ctype, contacts, phone, address, remark, cid);
        DBHelper db = new DBHelper();
        return db.Operate(sql);
    }
```

表示层代码

```
protected void Page_Load(object sender, EventArgs e)
    {//页面首次加载时,显示全部客户信息列表
        if (! IsPostBack)
        {
clientBLL tc = new clientBLL ();
            GridView1.DataSource = tc.getclient ("");
            GridView1.DataBind();
        }
    }
//按关键字查询客户信息
    protected void Button1_Click(object sender, EventArgs e)
    {
```

```
            clientBLL tc = new clientBLL();
        DataSet ds = tc.getclient(TextBox1.Text.Trim());
        if(ds.Tables[0].Rows.Count > 0)//如果查询结果集中的不为空
        {
            GridView1.DataSource = ds;
            GridView1.DataBind();
        }
//添加公司信息
protected void LinkButton1_Command(object sender, EventArgs e)
{
String cid = Request.QueryString["Cid"].ToString();
        String cname,ctype,contacts,phone,address,remark;
        cname = TextBox1.Text;
        ctype = DropDownList1.Text;
        contacts = TextBox2.Text;
    phone = TextBox3.Text;
    address = TextBox4.Text;
    remark = TextBox5.Text;
    clientBLL tc = new clientBLL();
    if(tc.updateclient(cid,cname,ctype,contacts,phone,address,remark) > 0)
        Response.Redirect("Main_client01.aspx");
}
```

任务十四：物流管理系统的车辆类型管理

一、任务描述

你作为《物流管理系统》项目开发组的程序员，请实现以下功能：
- 显示车辆类型列表；
- 查询车辆类型；
- 添加车辆类型。

二、功能描述

（1）点击图14.1所示页面顶部导航条中的"运输管理"菜单项，在打开的左侧菜单中点击"车辆类型"菜单项，则在右边的主体部分显示车辆类型信息列表。

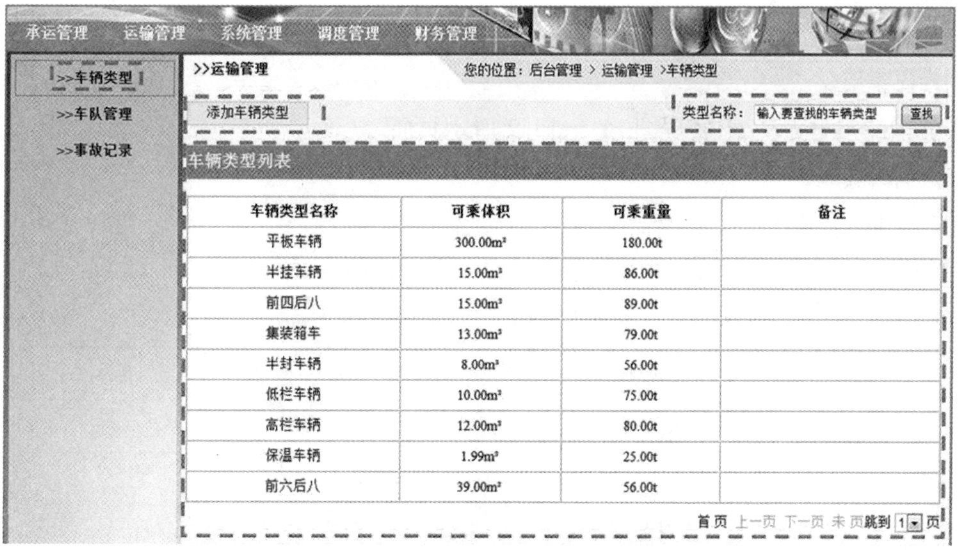

图14.1 车辆类型信息列表页面

（2）在图14.1中，输入需查询的车辆类型名称，点击"查找"按钮，将查询结果显示在车辆类型列表中。

(3)在图14.1中,点击"添加车辆类型"按钮,则跳转到车辆类型信息录入页面,如图14.2所示。

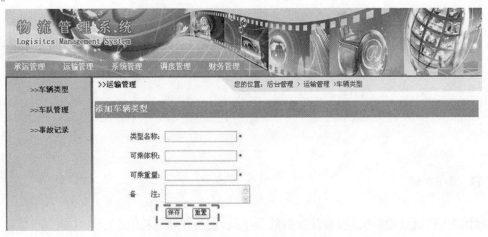

图14.2 公司信息录入页面

(4)点击图14.2中的"保存"按钮,对图中打"＊"号的输入部分进行必填校验,通过校验后在数据库中添加车辆类型信息。

(5)车辆类型信息增加成功后,跳转到图4.61所示页面,显示更新后的车辆类型信息列表。

(6)测试程序,在添加车辆类型信息页面新增两条以上车辆类型信息。

三、要求

1. 界面实现

以提供的素材为基础,实现图14.1、14.2所示页面。

2. 数据库实现

(1)创建数据库 LogisticsDB。

(2)创建车辆类型信息表(T_logistics_car_type),表结构见表14.1。

表14.1 车辆类型信息表(T_logistics_car_type)表结构

字段名	字段说明	字段类型	允许为空	备注
Cp_id	类型ID	varchar(10)	否	主键
Cp_name	类型名称	varchar(20)	否	
Cp_volume	可乘体积	float	否	单位为m^3
Cp_weight	可乘重量	float	否	单位为t
Cp_remark	备注	varchar(100)	是	

(3)在表 T_logistics_car_type 插入以下记录,见表14.2。

表 14.2　车辆类型信息表（T_logistics_car_type）记录

Cp_id	Cp_name	Cp_volume	Cp_weight	Cp_remark
CP0001	平板车辆	300.00	180.00	
CP0002	半挂车辆	15.00	86.00	
CP0003	前四后八	15.00	89.00	

3. 功能实现

（1）功能需求如图 14.3 所示。

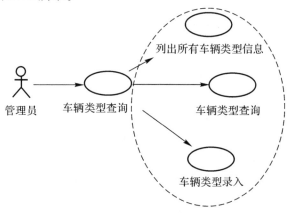

图 14.3　车辆类型模块用例图

（2）依据车辆类型列表活动图完成车辆类型信息列表显示功能，如图 14.4 所示。

（3）依据添加车辆类型信息活动图完成添加车辆类型信息功能，如图 14.5 所示。

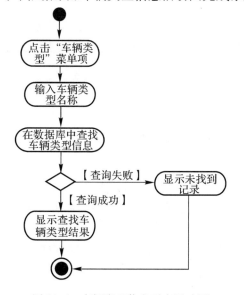

图 14.4　车辆类型信息列表活动图

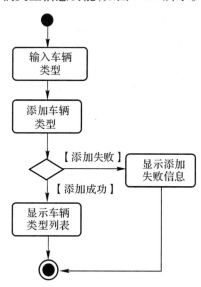

图 14.5　添加车辆类型信息活动图

四、必备知识

1. 数据库相关知识

（1）使用 MS SQL Server 2005/2008 创建数据库，创建数据表，设置表的字段，数据类型，主键，外键，约束。

（2）向数据表插入、删除、修改、查询数据。

2. 页面相关知识

（1）看懂系统所提供的素材，并理解框架技术。

（2）使用 asp.net 验证控件对页面必要的内容进行校验。

3. Asp.net 相关知识

（1）理解如何将 aspx 页面嵌入到框架中。

（2）使用数据源绑定控件显示数据库的数据。

（3）4.3.3 合理的使用转发和重定向控制项目的页面跳转。

（4）使用 ado.net 技术与数据库进行交互。

Asp.net 三层结构的数据访问层、业务逻辑层和表示层的功能和合作。

五、解题思路

1. 数据库思路

（1）根据项目要求创建数据库和数据表，向数据表中插入合适的测试数据。

（2）添加数据访问层，并在其中写好连接字符串和若干个通用的数据库操作方法。

（3）添加业务逻辑层，设置适当的参数，实现本任务所要求的全部功能，本层可以引用数据访问层的方法。

2. 表示层思路

（1）将提供的素材页面改写为 aspx 页面。

（2）将数据库数据的显示控件和部分表单控件改成 asp 服务器端控件。

3. 实现思路

（1）在表示层访问业务逻辑层，按照题目要求调用业务逻辑层的方法。

（2）数据添加的时候，正确获取用户在表单中提取的数据，并写入数据库中，然后进行页面跳转逻辑控制。

（3）查询的时候，调用业务逻辑层的方法，获取数据库中的数据，并按照用户要求的格式显示出来。

六、操作步骤

//查询全部或者部分车辆信息列表

```csharp
        public DataSet getcar(string key)
        {
            string sql = string.Format("select * from T_logistics_car_type where Cp_name like '%{0}%'", key);
            DBHelper db = new DBHelper();
            return db.FillDataSet(sql);
        }
    //添加公司信息
        public int addcar(string Cp_id, string Cp_name, string Cp_volume, string Cp_weight, string Cp_remark)
        {
            string sql = string.Format("insert into T_logistics_car_type (Cp_id ,Cp_name, Cp_volumne, Cp_weight, Cp_remark) values('{0}','{1}','{2}','{3}','{4}')", Cp_id, Cp_name, Cp_volume, Cp_weight, Cp_remark);
            DBHelper db = new DBHelper();
            return db.Operate(sql);
        }
```

表示层代码

```csharp
protected void Page_Load(object sender, EventArgs e)
    {//页面首次加载时,显示全部车辆信息
        if (!IsPostBack)
        {
            carBLL tc = new carBLL();
            GridView1.DataSource = tc.getcar("");
            GridView1.DataBind();
        }
    }
//按关键字查询车辆信息
    protected void Button1_Click(object sender, EventArgs e)
    {
            carBLL tc = new carBLL();
        DataSet ds = tc.getcar(TextBox1.Text.Trim());
        if (ds.Tables[0].Rows.Count > 0)//如果查询结果集中的不为空
        {
            GridView1.DataSource = ds;
            GridView1.DataBind();
        }
//添加车辆类型
protected void Button1_Click(object sender, EventArgs e)
    {
        if (Page.IsValid)
        {
            stringCp_id, Cp_name, Cp_volume, Cp_weight, Cp_remark;Cp_id = TextBox1.Text;
            Cp_name = TextBox2.Text;
```

```
            Cp_volume = TextBox3. Text;
            Cp _weight = TextBox4. Text;
            Cp_remark = TextBox5. Text;
            carBLL tc = new carBLL( );
            if ( tc. addcar ( Cp_id, Cp_name, Cp_volume, Cp _weight, Cp_remark) > 0)
                Response. Redirect( "Main_carriage01. aspx" );
        }
    }
```

任务十五：物流管理系统的车队信息管理

一、任务描述

你作为《物流管理系统》项目开发组的程序员，请实现如下功能：
- 显示车队信息列表；
- 查询车队信息；
- 删除车队信息。

二、功能描述

（1）点击图15.1所示页面顶部导航条中的"运输管理"菜单项，在打开的左侧菜单中点击"车队管理"菜单项，则在右边的主体部分显示车队信息列表。

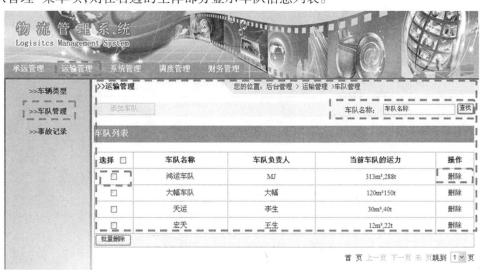

图15.1 车队信息列表页面

（2）在图15.1中，输入需查询的车队名称，点击"查找"按钮，将查询结果显示在车队信息列表中。

（3）在图15.1中，点击操作列中的"删除"按钮，弹出删除警告，点击"确定"后执行删除操作，删除所在行的客户信息。

(4)删除车队信息成功后,跳转到图 15.1 所示页面,显示更新后的车队信息列表。

三、要求

1. 界面实现

以提供的素材为基础,实现图 15.1 所示页面。

2. 数据库实现

(1)创建数据库 LogisticsDB。

(2)创建车队信息表(T_logistics_fleet),表结构见表 15.1。

表 15.1　车队信息表(T_logistics_fleet)表结构

字段名	字段说明	字段类型	允许为空	备注
Fleet_id	车队 ID	varchar(5)	否	主键
Fleet_name	车队名称	varchar(50)	否	
Fleet_functionary	负责人	varchar(20)	否	
Fleet_remark	备注	varchar(100)	是	

(3)在表 T_logistics_fleet 插入以下记录,见表 15.2。

表 15.2　车队信息表(T_logistics_fleet)记录

Fleet_id	Fleet_name	Fleet_functionary	Fleet_remark
F0001	鸿运车队	MJ	313m3,288t
F0002	大幅车队	大幅	120m3,150t
F0003	天运	李生	30m3,40t
F0004	宏天	王生	12m3,22t

3. 功能实现

(1)功能需求如图 15.2 所示。

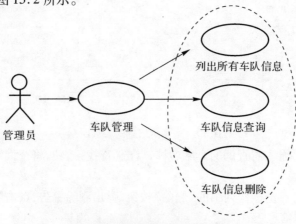

图 15.2　车队管理模块用例图

(2)依据车队信息列表活动图完成车队信息列表显示功能,如图 15.3 所示。
(3)依据删除车队信息活动图完成删除车队信息功能,如图 15.4 所示。

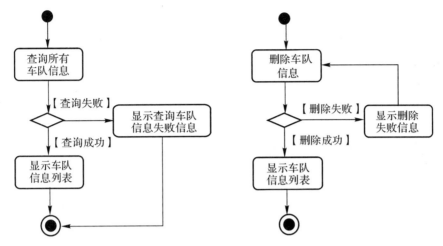

图 15.3　车队信息列表活动图　　　　图 15.4　删除车队信息活动图

四、必备知识

1. 数据库相关知识

(1)使用 MS SQL Server 2005/2008 创建数据库,创建数据表,设置表的字段,数据类型,主键,外键,约束。
(2)向数据表插入、删除、修改、查询数据。

2. 页面相关知识

(1)看懂系统所提供的素材,并理解框架技术。
(2)使用 asp.net 验证控件对页面必要的内容进行校验。

3. Asp.net 相关知识

(1)理解如何将 aspx 页面嵌入到框架中。
(2)使用数据源绑定控件显示数据库的数据。
(3)合理地使用转发和重定向控制项目的页面跳转。
(4)使用 ado.net 技术与数据库进行交互。
(5)Asp.net 三层结构的数据访问层、业务逻辑层和表示层的功能和合作。

五、解题思路

1. 数据库思路

(1)根据项目要求创建数据库和数据表,向数据表中插入合适的测试数据。
(2)添加数据访问层,并在其中写好连接字符串和若干个通用的数据库操作方法。
(3)添加业务逻辑层,设置适当的参数,实现本任务所要求的全部功能,本层可以引用数据访问层的方法。

2. 表示层思路

(1)将提供的素材页面改写为 aspx 页面。

(2)将数据库数据的显示控件和部分表单控件改成 asp 服务器端控件。

3. 实现思路

(1)在表示层访问业务逻辑层,按照题目要求调用业务逻辑层的方法。

(2)数据添加的时候,正确获取用户在表单中提取的数据,并写入数据库中,然后进行页面跳转逻辑控制。

(3)查询的时候,调用业务逻辑层的方法,获取数据库中的数据,并按照用户要求的格式显示出来。

六、操作步骤

```
//显示所有车队信息
public bool GetAllInformation(string Fleet_name)
    {
        string sql = string.Format("select * from T_ T_logistics_fleet");
        DBHelper db = new DBHelper();
        return db.FillDataSet(sql);
    }
//输入车队名称查找车队
public bool GetByName(string Fleet_name)
    {
        string sql = string.Format("select * from T_ T_logistics_fleet where Fleet_name='{0}'", Fleet_name);
        DBHelper db = new DBHelper();
        return db.FillDataSet(sql);
    }
//删除车队
    public bool delete(string Fleet_id)
    {
        string sql = string.Format("delete from T_logistics_fleet where Fleet_id='{0}'", Fleet_id);
        DBHelper db = new DBHelper();
        if (db.Operate(sql) > 0)
            return true;
        else
            return false;
    }
表示层
protected void Page_Load(object sender, EventArgs e)
    {
```

```csharp
        //页面加载事件,声明业务逻辑层对象,用于调用其中的方法
        BLL pb = new BLL();
        GridView1.DataSource = pb.GetAllInformation();
        GridView1.DataBind();
    }
//点击查找按钮执行的操作
    protected void Button1_Click(object sender, EventArgs e)
    {
        BLL pb = new BLL();
        GridView1.DataSource = pb.GetByName();
        GridView1.DataBind();
}
//点击删除按钮执行的操作
protected void Button2_Command(object sender, CommandEventArgs e)
{
        //如果当前页面通过验证
        if(IsValid)
        {
            BLL pb = new BLL();

            if(pb.delete(sql)>0)//如果信息添加成功,则跳转到项目工程查看页面
                Response.Redirect("Main_cariage03.aspx");
        }
}
```

任务十六：码头中心船货申报系统的船货作业信息管理

一、任务描述

你作为《码头中心船货申报系统》项目开发组的程序员，请实现如下功能：
- 显示船货作业信息列表；
- 添加船货作业信息。

二、功能描述

（1）点击图 16.1 所示页面顶部导航条中的"船货申报管理"菜单项，则在下部的主体部分显示船货作业信息列表。

图 16.1　船货作业信息列表页面

（2）在图 16.1 中，点击"增加申报"按钮，进入船货作业信息录入页面，如图 16.2 所示。

图 16.2　船货作业信息录入页面

（3）点击图 16.2 中的"确定"按钮，对图中打"＊"号的输入部分进行必填校验，通过校验后在数据库中添加船货作业信息。

(4)船货作业信息增加成功后,跳转到图16.1所示页面,显示更新后的船货作业信息列表。

(5)测试程序,在添加船货作业信息页面新增两条以上船货作业信息。

三、要求

1. 界面实现

以提供的素材为基础,实现图16.1、图16.2所示页面。

2. 数据库实现

(1)新建数据库HarborBureauDB。

(2)新建船货申报信息表(T_cargo_declare),表结构见16.1。

表16.1 船货申报信息表(T_cargo_declare)表结构

字段名	字段说明	字段类型	允许为空	备注
Declare_id	申报编号	varchar(50)	否	主键
Ship_name	船舶名称	varchar(50)	否	
Berth_name	泊位名称	varchar(50)	否	
Cargo	货物名称	varchar(50)	是	
Declare_ton	申报吨	int	是	

(3)在表T_cargo_declare插入以下记录,见表16.2。

表16.2 船货申报信息表(T_cargo_declare)记录

Declare_id	Ship_name	Berth_name	Cargo	Declare_ton
2008/08/212005	双拥号	12号泊位	矿石	35
2008/08/213256	友好号	12号泊位	矿石	56

3. 功能实现

(1)功能需求如图16.3所示。

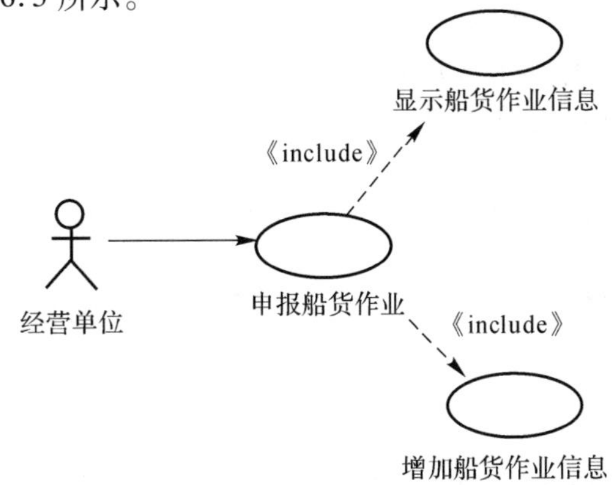

图16.3 船货作业申报模块用例图

(2)依据船货作业信息列表活动图完成船货作业信息列表显示功能,如图 16.4 所示。

(3)依据增加船货作业信息活动图完成增加船货作业信息功能,如图 16.5 所示。

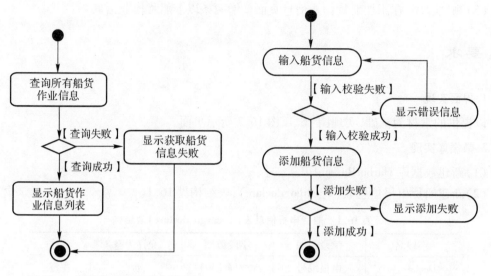

图 16.4　船货作业信息列表活动图　　图 16.5　添加船货作业信息活动图

四、必备知识

1. 数据库相关知识

(1)使用 MS SQL Server 2005/2008 创建数据库,创建数据表,设置表的字段,数据类型,主键,外键,约束。

(2)向数据表插入、删除、修改、查询数据。

2. 页面相关知识

(1)看懂系统所提供的素材,并理解框架技术。

(2)使用 asp.net 验证控件对页面必要的内容进行校验。

3. Asp.net 相关知识

(1)理解如何将 aspx 页面嵌入到框架中。

(2)使用数据源绑定控件显示数据库的数据。

(3)合理的使用转发和重定向控制项目的页面跳转。

(4)使用 ado.net 技术与数据库进行交互。

(6)Asp.net 三层结构的数据访问层、业务逻辑层和表示层的功能和合作

五、解题思路

1. 数据库思路

(1)根据项目要求创建数据库和数据表,向数据表中插入合适的测试数据。

(2)添加数据访问层,并在其中写好连接字符串和若干个通用的数据库操作方法。

(3)添加业务逻辑层,设置适当的参数,实现本任务所要求的全部功能,本层可以引用数据访问层的方法。

2. 表示层思路

(1)将提供的素材页面改写为 aspx 页面。

(2)将数据库数据的显示控件和部分表单控件改成 asp 服务器端控件。

3. 实现思路

(1)在表示层访问业务逻辑层,按照题目要求调用业务逻辑层的方法

(2)数据添加的时候,正确获取用户在表单中提取的数据,并写入数据库中,然后进行页面跳转逻辑控制。

(3)查询的时候,调用业务逻辑层的方法,获取数据库中的数据,并按照用户要求的格式显示出来。

六、操作步骤

业务逻辑层代码:

```
//显示船货作业信息列表;
public DataSet getallcargo( )
{ DBhelper db = new DBhelper( );
    string sql = "select * from T_cargo_declare";
    return db.FillDataSet(sql);
}
//添加船货作业信息。
public bool addcargo(string Declare_id, string Ship_name, string Berth_name, string Cargo, string Decalre_ton)
{
DBhelper db = new DBhelper( );
    string sql = string.Format("insert into T_cargo_declare values('{0}','{1}','{2}','{3}','{4}')",Declare_id,Ship_name,Berth_name,Cargo,Decalre_ton);
    if (db.Operate(sql) > 0)
        return true;
    else
        return false;
```

表示层代码:

```
protected void Page_Load(object sender, EventArgs e)
{///显示船货作业信息列表;
cargo  go = new cargo( );
    GridView1.DataSource = dp.getallcargo( );
    GridView1.DataBind( );
}
```

```csharp
//添加船货作业信息
        protected void Button1_Click(object sender, EventArgs e)
        {
String Declare_id,Ship_name,Berth_name,Cargo,Decalre_ton;
cargo   go = new cargo();
            Declare_id = TextBox1.Text.Trim();
            Ship_name = TextBox2.Text.Trim();
         Berth_name = TextBox3.Text;
         Cargo = TextBox4.Text;
         Decalre_ton = TextBox5.Text;
          bool ok = dp.addcargo(Declare_id,Ship_name,Berth_name,Cargo,Decalre_ton);
            if(ok)//如果信息添加成功,则显示更新后的船货作业信息列表
                Response.Redirect("Default.aspx");
        }
```

任务十七:码头中心船货申报系统的进出港旅客流量信息管理

一、任务描述

你作为《码头中心船货申报系统》项目开发组的程序员,请实现如下功能:
➤ 显示进出港旅客流量信息列表;
➤ 添加进出港旅客流量信息。

二、功能描述

(1)点击图17.1所示页面顶部导航条中的"船货申报管理"菜单中的"水上进出港旅客客流量列表"菜单项,则在中间的主体部分显示所有旅客客流量申报信息列表,如图17.1所示。

图17.1 水上进出港旅客客流量信息列表页面

(2)在图17.1中,点击"水上进出港旅客客流量申报"超链,则跳转到水上进出港旅客客流量申报页面,如图17.2所示。

(3)点击图17.2中的"确定"按钮,对图中打"＊"号的输入部分进行必填校验,通过校验后在数据库中添加旅客客流量信息。

(4)旅客客流量信息增加成功后,跳转到图17.1所示页面,显示更新后的旅客客流量信息列表。

(5)测试程序,在水上进出港旅客客流量申报页面新增两条以上旅客客流量信息。

图17.2 水上进出港旅客客流量申报页面

三、要求

1. 界面实现

以提供的素材为基础,实现图17.1、图17.2所示页面。

2. 数据库实现

(1)新建数据库 HarborBureauDB。

(2)新建旅客客流量申报表(T_guest_declare),表结构见表17.1。

表17.1 旅客客流量申报表(T_guest_declare)表结构

字段名	字段说明	字段类型	允许为空	备注
Declare_no	申报编号	nvarchar(12)	否	主键
Line_code	航线代码	nvarchar(5)	否	
Voyage_number_code	航次代码	nchar(10)	是	
In_out_port	进出港	nvarchar(4)	是	
Checked_guest_qty	核定载客数	int	是	
Declarer	申报人	nvarchar(24)	是	

(3)在表 T_guest_declare 插入以下记录,见表17.2。

表17.2 旅客客流量申报表(T_guest_declare)记录

Declare_no	Line_code	Voyage_number_code	In_out_port	Checked_guest_qty	Declarer
200110123456	H201	034K	进港	350	谢长江
200609087654	H101	015S	出港	300	吴江洪
200809083212	H201	034K	进港	350	谢长江

(4)航线名称的下拉列表框中的数据见表17.3。

表17.3　航线名称下拉列表框数据

航线编码	航线名称
H101 上海	桑给巴尔
H201 长兴	吴淞

显示在下拉列表框中是航线名称,插入到 T_guest_declare(旅客客流量申报表)表中的是航线编码。

(5)进出港下拉列表框中的数据为{"进港","出港"}。

3.功能实现。

(1)功能需求如图7.3所示。

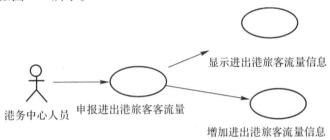

图17.3　进出港旅客流量模块用例图

(3)2 依据进出港旅客流量信息列表活动图完成进出港旅客流量信息列表显示功能,如图17.4所示。

(3)依据增加进出港旅客流量信息活动图完成增加进出港旅客流量信息功能,如图17.5所示。

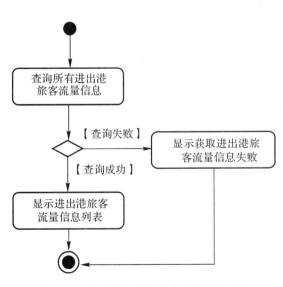

图17.4　进出港旅客流量信息列表活动图

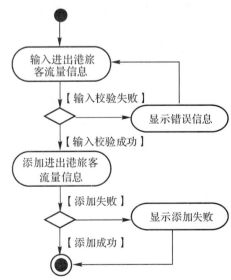

图17.5　增加进出港旅客流量信息活动图

四、必备知识

1. 数据库相关知识

（1）使用 MS SQL Server 2005/2008 创建数据库，创建数据表，设置表的字段，数据类型，主键，外键，约束。

（2）向数据表插入、删除、修改、查询数据。

2. 页面相关知识

（1）看懂系统所提供的素材，并理解框架技术。

（2）使用 asp.net 验证控件对页面必要的内容进行校验。

3. Asp.net 相关知识

（1）理解如何将 aspx 页面嵌入到框架中。

（2）使用数据源绑定控件显示数据库的数据。

（3）合理的使用转发和重定向控制项目的页面跳转。

（4）使用 ado.net 技术与数据库进行交互。

（5）Asp.net 三层结构的数据访问层、业务逻辑层和表示层的功能和合作。

五、解题思路

1. 数据库思路

（1）根据项目要求创建数据库和数据表，向数据表中插入合适的测试数据。

（2）添加数据访问层，并在其中写好连接字符串和若干个通用的数据库操作方法。

（3）添加业务逻辑层，设置适当的参数，实现本任务所要求的全部功能，本层可以引用数据访问层的方法。

2. 表示层思路

（1）将提供的素材页面改写为 aspx 页面。

（2）将数据库数据的显示控件和部分表单控件改成 asp 服务器端控件。

3. 实现思路

（1）在表示层访问业务逻辑层，按照题目要求调用业务逻辑层的方法。

（2）数据添加的时候，正确获取用户在表单中提取的数据，并写入数据库中，然后进行页面跳转逻辑控制。

（3）查询的时候，调用业务逻辑层的方法，获取数据库中的数据，并按照用户要求的格式显示出来。

六、操作步骤

业务逻辑层代码：

```csharp
//显示进出港旅客流量信息列表;
public DataSet getallguest()
{
    DBhelper db = new DBhelper();
    string sql = "select * fromT_guest_declare";
    return db.FillDataSet(sql);
}
//添加进出港旅客流量信息
public bool addguest(string Declare_no, string Line_code, string Voyage_number_code, string In_out_port, string Checked_guest_qty, string Declarer)
{
    DBhelper db = new DBhelper();
    string sql = string.Format("insert intoT_guest_declare values('{0}','{1}','{2}','{3}','{4}','{5}')", Declare_no, Line_code, Voyage_number_code, In_out_port, Checked_guest_qty, Declarer);
    if (db.Operate(sql) > 0)
        return true;
    else
        return false;
}
```

表示层代码:

```csharp
protected void Page_Load(object sender, EventArgs e)
{ ///显示进出港旅客流量信息列表;
    guestBLL gb = new guestBLL();
    GridView1.DataSource = dp.getallguest();
    GridView1.DataBind();
}
//添加进出港旅客流量信息
protected void Button1_Click(object sender, EventArgs e)
{
    String Declare_no, Line_code, Voyage_number_code, In_out_port, Checked_guest_qty, Declarer;
    guestBLL gb = new guestBLL();
    Declare_no = TextBox1.Text.Trim();
    Line_code = DropDownList1.Text;
    Voyage_number_code = TextBox2.Text.Trim();
    In_out_port = DropDownList2.Text;
    Checked_guest_qty = TextBox3.Text;
    Declarer = TextBox4.Text;
    bool ok = dp.addguest(Declare_no, Line_code, Voyage_number_code, In_out_port, Checked_guest_qty, Declarer);
    if (ok)//如果信息添加成功,则显示更新后的旅客客流量信息列表
        Response.Redirect("Default.aspx");
}
```

任务十八：易居房产信息网的楼盘信息管理系统报建申请模块

一、任务描述

你作为《易居房产信息网》项目开发组的程序员，请实现如下功能：
- 发布楼盘信息；
- 审核楼盘信息。

二、功能描述

（1）点击图18.1所示页面左边导航条中的"楼盘信息发布"，则在右边的主体部分显示楼盘信息发布页面。

图18.1 楼盘信息发布页面原型

（2）在图18.1中，输入楼盘信息，点击"提交"按钮，对图中打"＊"号的输入部分进行必填校验，通过校验后在数据库中添加楼盘信息。

（3）点击图18.1所示页面左边导航条中的"楼盘信息审核"，则在右边的主体部分显楼盘信息审核页面,如图 18.2 所示。

图 18.2 楼盘信息审核页面原型

(4)5 在图 13.8 中,点击操作列中的"审核通过"链接,修改该楼盘信息的状态为"审核通过",并重新跳转到楼盘信息审核页面,显示更新后的楼盘信息列表。

(5)测试程序,发布两条楼盘信息,并将其状态修改为"审核通过"。

三、要求

1. 界面实现

以提供的素材为基础,实现图 18.1、图 18.1 所示页面。

2. 数据库实现

(1)创建数据库 EstateDB。
(2)创建项目信息表(T_building),表结构见表 18.1。

表 18.1 项目信息表(T_building)表结构

字段名	字段说明	字段类型	允许为空	备注
Id	楼盘信息编号	varchar(20)	否	主键
Company	开发商	varchar(40)	否	
Phone	联系电话	varchar(20)	否	
Description	楼盘描述	varchar(200)	否	
Status	状态	varchar(8)	否	

(3)在表 T_building 插入以下记录,见表 18.2。

表 18.2　项目信息表(T_building)记录

Id	Company	Phone	Description	Status
001	恒大地产	0731-88859908	东湖花园,10.1 开盘	审核通过
002	旭日地产	0731-82285588	西子花园,10.1 开盘	未审核
003	万达地产	0731-84155688	万达花园,10.1 开盘	未审核

3. 功能实现

(1) 功能需求如图 18.3 所示。

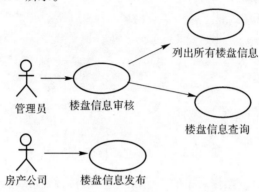

图 18.3　易居房产信息网楼盘发布模块用例图

(2) 依据楼盘信息发布活动图完成楼盘信息发布功能,如图 18.4 所示。

(3) 依据楼盘信息审核活动图完成楼盘信息审核功能,如图 18.5 所示。

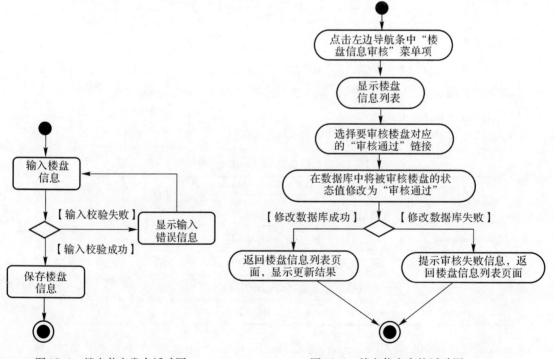

图 18.4　楼盘信息发布活动图　　图 18.5　楼盘信息审核活动图

四、必备知识

1. 数据库相关知识

（1）使用 MS SQL Server 2005/2008 创建数据库,创建数据表,设置表的字段,数据类型,主键,外键,约束。

（2）向数据表插入、删除、修改、查询数据。

2. 页面相关知识

（1）看懂系统所提供的素材,并理解框架技术。

（2）使用 asp.net 验证控件对页面必要的内容进行校验。

3. Asp.net 相关知识

（1）理解如何将 aspx 页面嵌入到框架中。

（2）使用数据源绑定控件显示数据库的数据。

（3）合理的使用转发和重定向控制项目的页面跳转。

（4）使用 ado.net 技术与数据库进行交互。

（5）Asp.net 三层结构的数据访问层、业务逻辑层和表示层的功能和合作。

五、解题思路

1. 数据库思路

（1）根据项目要求创建数据库和数据表,向数据表中插入合适的测试数据。

（2）添加数据访问层,并在其中写好连接字符串和若干个通用的数据库操作方法。

（3）添加业务逻辑层,设置适当的参数,实现本任务所要求的全部功能,本层可以引用数据访问层的方法。

2. 表示层思路

（1）将提供的素材页面改写为 aspx 页面。

（2）将数据库数据的显示控件和部分表单控件改成 asp 服务器端控件。

3. 实现思路

（1）在表示层访问业务逻辑层,按照题目要求调用业务逻辑层的方法。

（2）数据添加的时候,正确获取用户在表单中提取的数据,并写入数据库中,然后进行页面跳转逻辑控制。

（3）查询的时候,调用业务逻辑层的方法,获取数据库中的数据,并按照用户要求的格式显示出来。

六、操作步骤

业务逻辑层：

```csharp
//建设工程项目信息的添加
        public bool AddProject(string id, string company, string phone, string description)
        {
            string sql = string.Format("insert into T_building values ('{0}','{1}','{2}','{3}')", id, company, phone, description);
            DBHelper db = new DBHelper();
            if (db.Operate(sql) > 0)
                return true;
            else
                return false;
        }
        //修改审核状态
        public bool UpdateProject(string status)
        {
            string sql = string.Format("update T_building set status ='{0}'", status);
            DBHelper db = new DBHelper();
            if (db.Operate(sql) > 0)
                return true;
            else
                return false;
        }
//显示所有楼盘信息
        public DataSet GetAllProject()
        {
            string sql = "select * from T_project";
            DBHelper db = new DBHelper();
            return db.FillDataSet(sql);
        }
```

表示层
```csharp
protected void Button1_Click(object sender, EventArgs e)
        {
            string id, company, phone, description;
            id = TextBox1.Text.Trim();
            company = TextBox2.Text.Trim();
            phone = TextBox3.Text.Trim();
            description = TextBox4.Text.Trim();
            //如果当前页面通过验证
            if (IsValid)
            {
                BLL pb = new BLL();
                bool isok = pb.AddProject(id, company, phone, description);
                if (isok)//如果信息添加成功,则跳转到项目工程查看页面
                    Response.Redirect("BuildingList.aspx");
```

```
            }
        }
        protected void LinkButton1_Command(object sender, CommandEventArgs e)
        {
            //获取信息编号 ID
            string id = e.CommandArgument.ToString();
            //调用 BLL 层的 UpdateProject 验证函数,返回是否通过验证的布尔值和实例化的 id 对象
            if(id.UpdateProject(id,newid))
            {
                //添加 User 类型对象到 Session
                Session["id"] = newid;
                Response.Redirect("BuildingReview.aspx");
            }
        }
//显示所有建设工程项目
protected void Page_Load(object sender, EventArgs e)
        {
//页面加载事件,声明业务逻辑层对象,用于调用其中的方法
            BLL pb = new BLL();
            GridView1.DataSource = pb.GetAllproject();
            GridView1.DataBind();
        }
```

任务十九：易居房产信息网的会员管理

一、任务描述

你作为《易居房产信息网》项目开发组的程序员，请实现如下功能：
- 会员注册；
- 会员审核。

二、功能描述

（1）点击图19.1所示页面左边导航条中的"会员注册"，则在右边的主体部分显示会员注册页面。

图19.1 会员注册页面

（2）在图19.1中，输入会员注册信息，对打"＊"号的输入部分进行必填校验，通过校验后在数据库中添加会员信息，其会员状态默认为"未审核"。

（3）点击图19.2所示页面左边导航条中的"会员审核"，进入会员审核页面。

图 19.2 会员审核页面

(4)在图 19.2 中,点击操作列中的"审核通过"链接,修改该会员信息的状态为"审核通过",并重新跳转到会员信息审核页面,显示更新后的会员信息列表。

(5)测试程序,注册两个以上用户,并将其状态修改为"审核通过"。

三、要 求

1. 界面实现

以提供的素材为基础,实现图 19.1、图 19.2 所示页面。

2. 数据库实现

(1)创建数据库 EstateDB。

(2)创建会员信息表(T_member),表结构见表 19.1。

表 19.1　会员信息表(T_member)表结构

字段名	字段说明	字段类型	允许为空	备注
Id	会员编号	varchar(20)	否	主键
Name	姓名	varchar(20)	否	
Phone	电话	varchar(20)	是	
Address	地址	varchar(20)	是	
Status	状态	varchar(8)	否	

(3)在表 T_member 插入以下记录,见表 19.2。

表 19.2 项目信息表(T_member)记录

Id	Name	Phone	Address	Status
001	王明	0731-28855990	建设路175号	审核通过
002	周文	0731-28854223	长岭路15号	未审核
003	刘伟	0731-82255996	长岭路157号	未审核

3. 功能实现

(1)功能需求如图 19.3 所示。

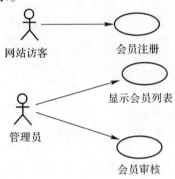

图 19.3 易居房产信息网会员管理模块用例图

(2)依据会员注册活动图完成会员注册功能,如图 19.4 所示。

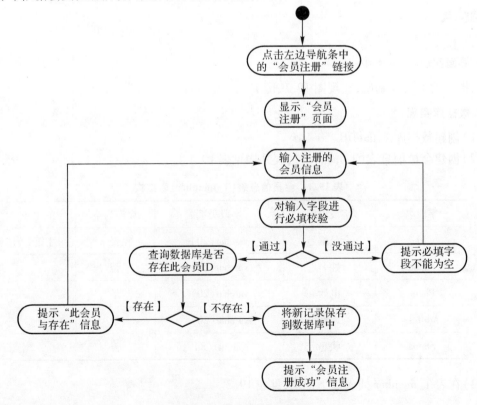

图 19.4 会员注册活动图

（3）依据会员审核活动图完成会员审核功能，如图 19.5 所示。

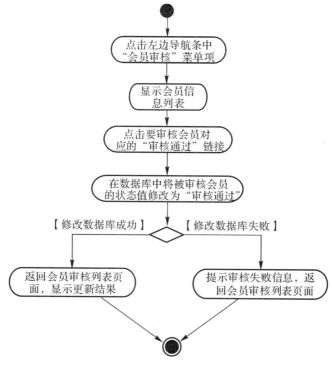

图 19.5　会员审核活动图

四、必备知识

1. 数据库相关知识

（1）使用 MS SQL Server 2005/2008 创建数据库，创建数据表，设置表的字段，数据类型，主键，外键，约束。

（2）向数据表插入、删除、修改、查询数据。

2. 页面相关知识

（1）看懂系统所提供的素材，并理解框架技术。

（2）使用 asp.net 验证控件对页面必要的内容进行校验。

3. Asp.net 相关知识

（1）理解如何将 aspx 页面嵌入到框架中。

（2）使用数据源绑定控件显示数据库的数据。

（3）合理的使用转发和重定向控制项目的页面跳转。

（4）使用 ado.net 技术与数据库进行交互。

（5）Asp.net 三层结构的数据访问层、业务逻辑层和表示层的功能和合作

五、解题思路

1. 数据库思路

（1）根据项目要求创建数据库和数据表，向数据表中插入合适的测试数据。

（2）添加数据访问层，并在其中写好连接字符串和若干个通用的数据库操作方法。

（3）添加业务逻辑层，设置适当的参数，实现本任务所要求的全部功能，本层可以引用数据访问层的方法。

2. 表示层思路

（1）将提供的素材页面改写为 aspx 页面。

（2）将数据库数据的显示控件和部分表单控件改成 asp 服务器端控件。

3. 实现思路

（1）在表示层访问业务逻辑层，按照题目要求调用业务逻辑层的方法。

（2）数据添加的时候，正确获取用户在表单中提取的数据，并写入数据库中，然后进行页面跳转逻辑控制。

（3）查询的时候，调用业务逻辑层的方法，获取数据库中的数据，并按照用户要求的格式显示出来。

六、操作步骤

业务逻辑层：

```
//会员注册
        public bool ZhuCe(string id,string name,string phone,string address)
    {
            string sql = string.Format("insert into T_member values ('{0}','{1}','{2}','{3}')",id, name, phone, address);
            DBHelper db = new DBHelper();
            if (db.Operate(sql) > 0)
                return true;
            else
                return false;
    }
        //显示所有会员信息
        public DataSet getallinformation()
    {
            string sql = "select * from T_member";
            DBHelper db = new DBHelper();
            return db.FillDataSet(sql);
    }
```

表示层

```csharp
//会员注册页面
protected void Button1_Click(object sender, EventArgs e)
{
    string id, name, phone, address;
    id = TextBox1.Text.Trim();
    name = TextBox2.Text.Trim();
    phone = TextBox3.Text.Trim();
    address = TextBox4.Text.Trim();
    if (IsValid)//如果当前页面通过验证
    {
        BLL pb = new BLL();
        bool isok = pb.ZhuCe(id, name, phone, address);
        if (isok)//如果信息添加成功,则跳转到项目工程查看页面
            Response.Redirect("MemberList.aspx");
    }
}
//显示所有会员信息
protected void Page_Load(object sender, EventArgs e)
{
    //页面加载事件,声明业务逻辑层对象,用于调用其中的方法
    BLL pb = new BLL();
    GridView1.DataSource = pb.getallinformation();
    GridView1.DataBind();
}
```

任务二十：易居房产信息网的房产出租信息管理

一、任务描述

你作为《易居房产信息网》项目开发组的程序员，请实现如下功能：
- 发布房产出租信息；
- 管理房产出租信息。

二、功能描述

（1）点击图20.1所示页面左边导航条中的"房产出租"，则在右边的主体部分显示房产出租信息发布页面。

图20.1 房产出租信息发布页面

（2）在图20.1中，输入房产出租信息，对打"＊"号的输入部分进行必填校验，通过校验后在数据库中添加房产出租信息。

(3)点击图20.2所示页面左边导航条中的"出租信息管理"链接,进入出租信息管理页面。

图20.2　房产出租信息删除页面

(4)在图20.2中,点击操作列中的"删除"链接,删除该出租信息,并重新跳转到房产出租信息列表页面,显示更新后的房产出租信息列表。

(5)测试程序,发布3条出租信息,并通过出租信息管理页面删除其中1条出租信息。

三、要求

1. 界面实现

以提供的素材为基础,实现图20.1、图20.2所示页面。

2. 数据库实现

(1)创建数据库 EstateDB。

(2)创建出租信息表(T_lease),表结构见表20.1。

表20.1　出租信息表(T_lease)表结构

字段名	字段说明	字段类型	允许为空	备注
Id	出租信息编号	varchar(20)	否	主键
Name	出租人	varchar(20)	否	
Phone	联系电话	varchar(20)	否	
Description	房产信息描述	varchar(200)	否	
Price	价格	double	否	

(3)在表 T_lease 插入以下记录,见表20.2。

表 20.2　出租信息表(T_lease)表记录

Id	Name	Phone	Description	Price
001	王明	073128855990	东湖花园,二楼,三室两厅	1500.0
002	周文	073128854223	东湖花园,一楼,一室一厅	800.0
003	刘伟	073182255996	万达花园,二楼,三室两厅	1800.0

3. 功能实现

(1) 功能需求如图 20.3 所示。

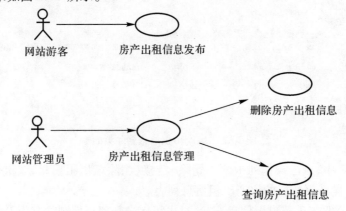

图 20.3　易居房产信息网房产出租模块用例图

(2) 依据出租信息发布活动图完成会员注册功能,如图 20.4 所示。

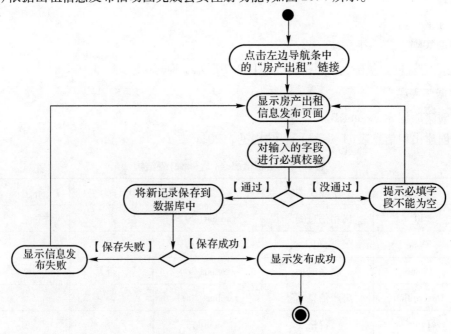

图 20.4　出租信息发布活动图

(3) 依据出租信息管理活动图完成出租信息删除功能,如图 20.5 所示。

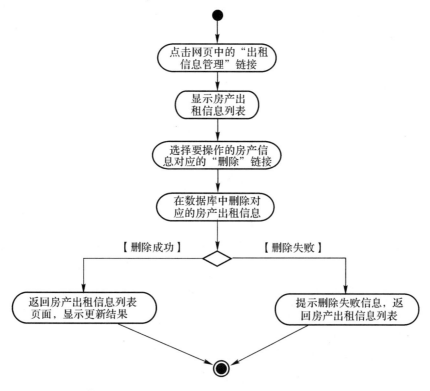

图 20.5　出租信息管理活动图

四、必备知识

1. 数据库相关知识

（1）使用 MS SQL Server 2005/2008 创建数据库，创建数据表，设置表的字段，数据类型，主键，外键，约束。

（2）向数据表插入、删除、修改、查询数据。

2. 页面相关知识

（1）看懂系统所提供的素材，并理解框架技术。

（2）使用 asp.net 验证控件对页面必要的内容进行校验。

3. Asp.net 相关知识

（1）理解如何将 aspx 页面嵌入到框架中。

（2）使用数据源绑定控件显示数据库的数据。

（3）合理的使用转发和重定向控制项目的页面跳转。

（4）使用 ado.net 技术与数据库进行交互。

（5）Asp.net 三层结构的数据访问层、业务逻辑层和表示层的功能和合作。

五、解题思路

1. 数据库思路

（1）根据项目要求创建数据库和数据表，向数据表中插入合适的测试数据。

（2）添加数据访问层，并在其中写好连接字符串和若干个通用的数据库操作方法。

（3）添加业务逻辑层，设置适当的参数，实现本任务所要求的全部功能，本层可以引用数据访问层的方法。

2. 表示层思路

（1）将提供的素材页面改写为 aspx 页面。

（2）将数据库数据的显示控件和部分表单控件改成 asp 服务器端控件。

3. 实现思路

（1）在表示层访问业务逻辑层，按照题目要求调用业务逻辑层的方法。

（2）数据添加的时候，正确获取用户在表单中提取的数据，并写入数据库中，然后进行页面跳转逻辑控制。

（3）查询的时候，调用业务逻辑层的方法，获取数据库中的数据，并按照用户要求的格式显示出来。

六、操作步骤

1. 业务逻辑层代码：

```
//出租信息管理显示
    public DataSet getallBuilding( )
    {
        string sql = " select * from T_building";
        DBHelper db = new DBHelper( );
        return db.FillDataSet(sql);
    }
//出租信息的发布
public bool AddBuilding(string id, string company, string phone, string description, string status)
{
    string sql = string.Format("insert into T_building values ('{0}','{1}','{2}','{3}','{4}')", id, company, phone, description, status);
    DBHelper db = new DBHelper( );
    if (db.Operate(sql) > 0)
        return true;
    else
        return false;
}
```

3. 表示层代码:

```
//出租信息管理显示
protected void Page_Load(object sender, EventArgs e)
    {// LeaseManage 页面加载事件
        BuildingBLL pb = new BuildingBLL();//声明业务逻辑层对象,用于调用其中的方法
        GridView1.DataSource = pb.getallBuilding();
        GridView1.DataBind();
    }
    //出租信息的添加
protected void Button1_Click(object sender, EventArgs e)
{
    String  id,company,phone,description,status;
    id = TextBox1.Text.Trim();
        company = TextBox2.Text;
        phone = TextBox3.Text;
        description = TextBox4.Text;
Status = TextBox5.Text;
if (IsValid)//如果当前页面通过验证
            {
                BuildingBLL pb = new BuildingBLL();
                bool isok = pb.Add Building(id,company,phone,description,status);
                if (isok)//如果信息添加成功,则跳转到出租信息页面
                    Response.Redirect("LeaseManage.aspx");
            }
```

任务二十一：教研室管理系统的个人信息管理

一、任务描述

你作为《教研室管理系统》项目开发组的程序员，请实现如下功能：
- 显示个人信息的列表；
- 添加个人信息。

二、功能描述

（1）点击图21.1所示页面左边导航条中的"个人资料管理"，则在右边的主体部分显示个人信息列表。

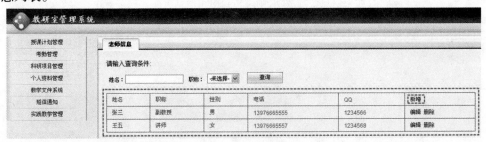

图21.1　个人信息列表页面

（2）在图21.1中，点击"新增"按钮，则跳转到个人信息录入页面，如图21.2所示。

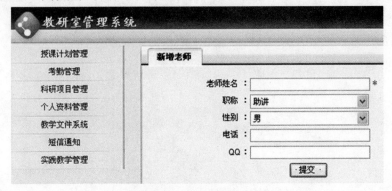

图21.2　个人信息录入页面

(3)在图21.2中,输入个人信息,对打"*"号的输入部分进行必填校验,通过校验后在数据库中添加个人信息。

(4)个人信息增加成功后,跳转到图21.1所示页面,显示更新后的个人信息列表。

(5)测试程序,增加两条以上个人信息。

三、要求

1. 界面实现

以提供的素材为基础,实现图21.1、图21.2所示页面。

2. 数据库实现

(1)创建数据库 TrsectionDB

(2)创建教师表(T_teacher),表结构见表21.1。

表21.1 教师表(T_teacher)表结构

字段名	字段说明	字段类型	允许为空	备注
Teacher_id	老师编号	int	否	主键
Teacher_name	老师姓名	varchar(20)	否	
Teacher_title	职称	varchar(20)	是	
Teacher_sex	性别	char(6)	是	
Telephone	电话	varchar(16)	是	
QQ	QQ	varchar(12)	是	

(3)在表T_teacher插入以下记录,见表21.2。

表21.2 T_teacher 表记录

Teacher_id	Teacher_name	Teacher_title	Teacher_sex	Telephone	QQ
1	张三	副教授	男	13976665555	123456
2	王五	讲师	女	13976665556	123458

3. 功能实现

(1)功能需求如图21.3所示。

(2)依据个人信息列表活动图完成个人信息列表显示功能,如图21.4所示。

(3)依据添加个人信息活动图完成添加个人信息功能,如图21.5所示。

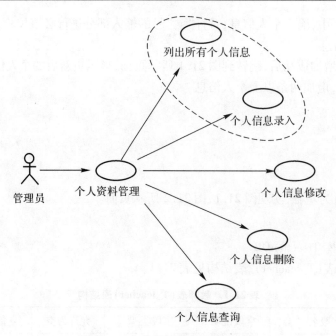

图 21.3 个人资料管理模块用例图

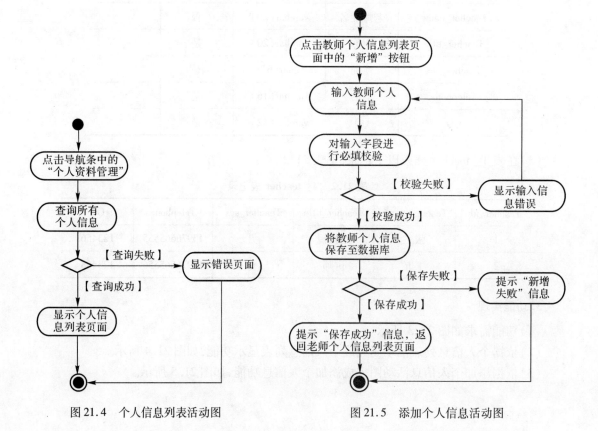

图 21.4 个人信息列表活动图　　　图 21.5 添加个人信息活动图

四、必备知识

1. 数据库相关知识

（1）使用 MS SQL Server 2005/2008 创建数据库，创建数据表，设置表的字段，数据类型，主键，外键，约束。

（2）向数据表插入、删除、修改、查询数据。

2. 页面相关知识

（1）看懂系统所提供的素材，并理解框架技术。

（2）使用 asp.net 验证控件对页面必要的内容进行校验。

3. Asp.net 相关知识

（1）理解如何将 aspx 页面嵌入到框架中。

（2）使用数据源绑定控件显示数据库的数据。

（3）合理的使用转发和重定向控制项目的页面跳转。

（4）使用 ado.net 技术与数据库进行交互。

（5）Asp.net 三层结构的数据访问层、业务逻辑层和表示层的功能和合作

五、解题思路

1. 数据库思路

（1）根据项目要求创建数据库和数据表，向数据表中插入合适的测试数据。

（2）添加数据访问层，并在其中写好连接字符串和若干个通用的数据库操作方法。

（3）添加业务逻辑层，设置适当的参数，实现本任务所要求的全部功能，本层可以引用数据访问层的方法。

2. 表示层思路

（1）将提供的素材页面改写为 aspx 页面。

（2）将数据库数据的显示控件和部分表单控件改成 asp 服务器端控件。

3. 实现思路

（1）在表示层访问业务逻辑层，按照题目要求调用业务逻辑层的方法。

（2）数据添加的时候，正确获取用户在表单中提取的数据，并写入数据库中，然后进行页面跳转逻辑控制。

（3）查询的时候，调用业务逻辑层的方法，获取数据库中的数据，并按照用户要求的格式显示出来。

六、操作步骤

1. 业务逻辑层代码

```csharp
//教师信息的列表显示
    public DataSet getallTeach()
    {
        string sql = "select * from T_teacher";
        DBHelper db = new DBHelper();
        return db.FillDataSet(sql);
    }
//显示想要查询的老师
public DataSet getTeach()
    {
        string sql = "select * from T_teacher where Teacher_name = 'TextBox1.Text' and Teacher_title = 'TextBox2.Text'";
        DBHelper db = new DBHelper();
        return db.FillDataSet(sql);
    }
//教师信息的添加
    public bool AddTeach(string Teacher_name,, string Teacher_title, string Teacher_sex, string Telephone, string QQ)
    {
        string sql = string.Format("insert into T_teacher values ('{0}','{1}','{2}','{3}','{4}')",Teacher_name,Teacher_title,Teacher_sex,Telephone,QQ);
        DBHelper db = new DBHelper();
        if(db.Operate(sql) > 0)
            return true;
        else
            return false;
    }
```

2. 表示层代码

```csharp
//老师信息的列表显示
protected void Page_Load(object sender, EventArgs e)
    {///teacherList 页面加载事件
        TeachBLL pb = new TeachBLL();//声明业务逻辑层对象,用于调用其中的方法
        GridView1.DataSource = pb.getTeach();
        GridView1.DataBind();
    }
//教师信息的添加
protected void Button1_Click(object sender, EventArgs e)
    {
```

```
String   Teacher_name,Teacher_title,Teacher_sex,Telephone,QQ;
Teacher_name = TextBox1.Text.Trim();
         Teacher_title = TextBox2.Text;
Teacher_sex = DropDownList1.SelectedValue;
   Telephone = TextBox4.Text;
         QQ = = TextBox5.Text;
if(IsValid)//如果当前页面通过验证
         {
             TeachBLL   pb = new TeachBLL();
              Boolv isok = pb.AddTeach(Teacher_name,Teacher_title,Teacher_sex,Telephone,QQ);
                 if(isok)//如果信息添加成功,则跳转到教师信息页面
                      Response.Redirect("teacherList.aspx");
         }
//筛选符合条件的老师
protected void Button1_Click(object sender, EventArgs e)
{
//teacherList 页面加载事件
         TeachBLL pb = new TeachBLL();//声明业务逻辑层对象,用于调用其中的方法

         GridView1.DataSource = pb.getTeach();
         GridView1.DataBind();
}
```

任务二十二:教研室管理系统的考勤功能

一、任务描述

你作为承接《教研室管理系统》项目开发组的程序员,请完成:
➢ 考勤管理的考勤功能。

二、功能描述

(1)在图 22.1 中点击教研室管理系统左边导航条中的"考勤管理",然后点击右边主体页面中的"输入考勤"按钮,系统会跳转到如图 22.2 所示的页面。

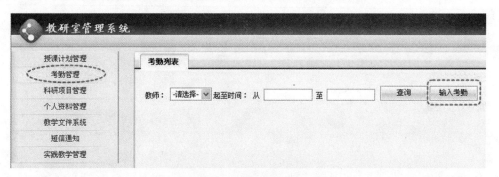

图 22.1　考勤管理页面

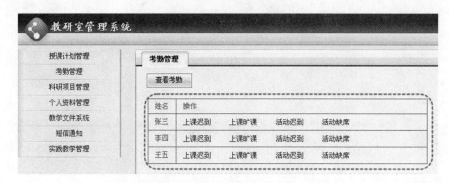

图 22.2　个人考勤页面

（2）在个人考勤页面中点击"上课迟到"、"上课旷课"、"活动迟到"、"活动缺席"中的某个按钮,将向考勤表中增加一条信息。

三、要求

1. 界面实现

以提供的素材为基础,实现图 22.1、图 22.2 所示页面。

2. 数据库实现

（1）创建数据库 TrsectionDB。

（2）创建教师信息表 T_teacher,表结构见表 22.1。

表 22.1 教师信息表（T_teacher）表结构

字段名	字段说明	字段类型	允许为空	备注
Teacher_id	老师编号	int	否	主键
Teacher_name	老师名称	varchar(20)	否	
Teacher_title	职称	varchar(20)	是	
Telephone	电话	varchar(16)	是	
QQ	QQ	varchar(12)	是	

（3）在表 T_teacher 插入以下记录,见表 22.2。

表 22.2 教师信息表（T_teacher）记录

Teacher_id	Teacher_name	Teacher_title	Telephone	QQ
1	张三	副教授	13976665555	123456
2	李四	讲师	13976665556	123457
3	王五	讲师	13976665557	123458

（4）创建考勤表 T_attendance,表结构见表 22.3。

表 22.3 考勤表（T_attendance）表结构

字段名	字段说明	字段类型	允许为空	备注
Attend_id	编号	int	否	主键
Attend_name	考勤名称	varchar(20)	否	"上课迟到"、"上课旷课"、"活动迟到"、"活动缺席"四种
Teacher_id	老师名称	int	否	外键
Attend_time	考勤时间	datetime	否	默认值为系统当前时间

（5）在表 T_attendance 插入以下记录,见表 22.4。

表 22.4 考勤表(T_attendance)表记录

Attend_id	Attend_name	Teacher_id	Attend_time
1	上课迟到	2	2011－5－27

3. 功能实现

(1)功能需求如图 22.3 所示。

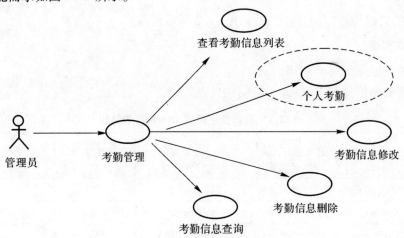

图 22.3 考勤管理模块用例图

(2)依据个人考勤活动图完成个人考勤功能,如图 22.4 所示。

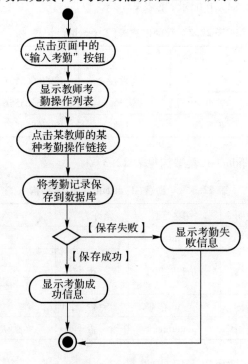

图 22.4 个人考勤活动图

四、必备知识

1. 数据库相关知识

（1）使用 MS SQL Server 2005/2008 创建数据库，创建数据表，设置表的字段，数据类型，主键，外键，约束。

（2）向数据表插入、删除、修改、查询数据。

2. 页面相关知识

（1）看懂系统所提供的素材，并理解框架技术。

（2）使用 asp.net 验证控件对页面必要的内容进行校验。

3. Asp.net 相关知识

（1）理解如何将 aspx 页面嵌入到框架中。

（2）使用数据源绑定控件显示数据库的数据。

（3）合理地使用转发和重定向控制项目的页面跳转。

（4）使用 ado.net 技术与数据库进行交互。

（5）Asp.net 三层结构的数据访问层、业务逻辑层和表示层的功能和合作。

五、解题思路

1. 数据库思路

（1）根据项目要求创建数据库和数据表，向数据表中插入合适的测试数据。

（2）添加数据访问层，并在其中写好连接字符串和若干个通用的数据库操作方法。

（3）添加业务逻辑层，设置适当的参数，实现本任务所要求的全部功能，本层可以引用数据访问层的方法。

2. 表示层思路

（1）将提供的素材页面改写为 aspx 页面。

（2）将数据库数据的显示控件和部分表单控件改成 asp 服务器端控件。

3. 实现思路

（1）在表示层访问业务逻辑层，按照题目要求调用业务逻辑层的方法。

（2）数据添加的时候，正确获取用户在表单中提取的数据，并写入数据库中，然后进行页面跳转逻辑控制。

（3）查询的时候，调用业务逻辑层的方法，获取数据库中的数据，并按照用户要求的格式显示出来。

六、操作步骤

1. 业务逻辑层代码

　　//对应老师表的所有操作

```csharp
public class T_teacherBLL
{
    public DataSet getallteacher()
    {
        string sql = "select * from T_teacher";
        DBHelper db = new DBHelper();
        return db.FillDataSet(sql);
    }
}
//对应考勤表的操作
public class T_attendanceBLL
{
    public int addattendance(string teacher_id, string Attend_name)
    {
        string sql = string.Format("insert into T_attendance(Attend_name,Teacher_id,Attend_time) values ('{0}',{1},getdate())", Attend_name, teacher_id);
        DBHelper db = new DBHelper();
        return db.Operate(sql);
    }
}
```

2. 表示层代码

```csharp
public partial class attendanceMain : System.Web.UI.Page
{
    protected void Page_Load(object sender, EventArgs e)
    {//将所有老师的姓名显示在页面上
        T_teacherBLL tb = new T_teacherBLL();
        GridView1.DataSource = tb.getallteacher();
        GridView1.DataBind();
    }
    protected void LinkButton1_Command(object sender, CommandEventArgs e)
    {//多控件共享单一事件,
        string id = e.CommandArgument.ToString();
        LinkButton lb = (LinkButton)sender;
        string command = lb.Text;
        T_attendanceBLL tb = new T_attendanceBLL();
        if (tb.addattendance(id, command) > 0)
            Response.Write("<script>alert('操作成功');</script>");
    }
}
```

任务二十三:银行信贷管理系统的保证金管理

一、任务描述

你作为《银行信贷管理系统》项目组的程序员,请实现如下功能:
➢ 添加保证金信息;
➢ 删除保证金信息。

二、功能描述

(1)在图23.1所示的银行信贷管理系统主页面中,点击左侧导航菜单"客户保证金管理"下的"添加保证金信息"链接,显示"添加保证金信息"页面,如图23.2所示。

图23.1 银行信贷管理系统主页面

(2)在图23.2中,录入保证金信息,"*"号表示该信息为必填字段,点击"保存信息"按钮,保存成功后跳转到"查询/编辑保证金信息"页面,如图23.3所示。
(3)在图23.3中,点击选定行的"删除"按钮,删除成功后跳转到"查询/编辑保证金信息"页面。

图23.2 添加客户保证金信息页面

图23.3 查询/编辑保证金信息页面

三、要求

1. 界面实现

以提供的素材为基础，实现图23.1～图23.3所示页面。说明如下：

在图23.2中，客户名称对应的下拉列表的value和Text属性用（2011070101,长沙创新科技有限公司），（2011070102,长沙奔流信息有限公司），（2011070103,长沙蓝海科技有限公司）三组值填充；

在图23.2中，冻结标志对应的下拉列表的value和Text属性用（Y,是）和（N,否）二组值填充；

在图23.2中，保证金状态对应的下拉列表的value和Text属性用（缴付，缴付）和（退还，退还)二组值填充。

2. 数据库实现

(1)创建数据库 BankCreditLoanDB。

(2)创建保证金信息表 T_bail_info,表结构见表 23.1。

表 23.1 保证金信息表(T_bail_info)表结构

字段名	字段说明	字段类型	是否允许为空	备注
Contract_id	合同号	CHAR(10)	否	主键,由 YYYYMMDD + 两位序号组成
Cust_id	客户编号	CHAR(10)	否	
Bail_account	保证金账号	CHAR(19)	否	
Bail_total_amount	保证金金额	DECIMAL(12,2)	否	
Bail_status	保证金状态	CHAR(4)	否	
FreezeFlag	冻结标志	CHAR(1)	否	Y 是 N 否

(3)在表 T_bail_info 中插入记录,见表 23.2。

表 23.2 保证金信息表(T_bail_info)记录

Contract_id	Cust_id	Bail_account	Bail_total_amount	Bail_status	FreezeFlag
2011070801	2011070101	6227 0000 1351 0065 598	100000	缴付	Y
2011070802	2011070102	6227 0000 1351 0065 599	200000	退还	N

3. 功能实现

(1)功能需求如图 23.4 所示。

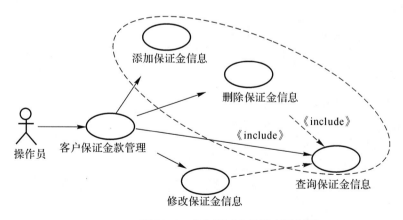

图 23.4 客户保证金款管理用例图

(2)依据活动图完成添加保证金信息功能,如图 23.5 所示。

(3)依据活动图完成删除保证金信息功能,如图 23.6 所示。

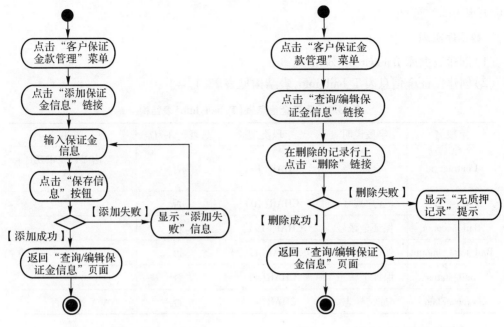

图 23.5　添加保证金信息活动图　　　图 23.6　删除保证金信息活动图

四、必备知识

1. 数据库相关知识

(1) 使用 MS SQL Server 2005/2008 创建数据库,创建数据表,设置表的字段,数据类型,主键,外键,约束。

(2) 向数据表插入、删除、修改、查询数据。

2. 页面相关知识

(1) 看懂系统所提供的素材,并理解框架技术。

(2) 使用 asp.net 验证控件对页面必要的内容进行校验。

3. Asp.net 相关知识

(1) 理解如何将 aspx 页面嵌入到框架中。

(2) 使用数据源绑定控件显示数据库的数据。

(3) 合理的使用转发和重定向控制项目的页面跳转。

(4) 使用 ado.net 技术与数据库进行交互。

(5) Asp.net 三层结构的数据访问层、业务逻辑层和表示层的功能和合作。

五、解题思路

1. 数据库思路

(1) 根据项目要求创建数据库和数据表,向数据表中插入合适的测试数据。

(2)添加数据访问层,并在其中写好连接字符串和若干个通用的数据库操作方法。

(3)添加业务逻辑层,设置适当的参数,实现本任务所要求的全部功能,本层可以引用数据访问层的方法。

2. 表示层思路

(1)将提供的素材页面改写为 aspx 页面。

(2)将数据库数据的显示控件和部分表单控件改成 asp 服务器端控件。

3. 实现思路

(1)在表示层访问业务逻辑层,按照题目要求调用业务逻辑层的方法。

(2)数据添加的时候,正确获取用户在表单中提取的数据,并写入数据库中,然后进行页面跳转逻辑控制。

(3)查询的时候,调用业务逻辑层的方法,获取数据库中的数据,并按照用户要求的格式显示出来。

六、操作步骤

1. 业务逻辑层代码

```
//保证金信息的列表显示
    public DataSet getallBail()
    {
        string sql = "select * from T_bail_info ";
        DBHelper db = new DBHelper();
        return db.FillDataSet(sql);
    }
//显示想要查询的信息
public DataSet get Bail()
    {
        string sql = "select * from T_bail_info where Cust_id = ' TextBox1.Text'";
        DBHelper db = new DBHelper();
        return db.FillDataSet(sql);
    }
    //添加保证金信息的添加
    public bool Add Bail (string Cust_id, string Bail_account, string Bail_total_amount, string Bail_status, string FreezeFlag)
    {
        string sql = string.Format("insert into T_bail_info values ('{0}','{1}','{2}','{3}','{4}')", Cust_id, Bail_account, Bail_total_amount, Bail_total_amount, Bail_status, FreezeFlag);
        DBHelper db = new DBHelper();
        if (db.Operate(sql) > 0)
            return true;
        else
```

```
            return false;
        }
}
```

2. 表示层代码

```
//保证金信息的列表显示
protected void Page_Load(object sender, EventArgs e)
    {// BailList 页面加载事件
        BailBLL pb = new BailBLL();//声明业务逻辑层对象,用于调用其中的方法
        GridView1.DataSource = pb.getallBail();
        GridView1.DataBind();
    }
    //保证金信息的添加
protected void Button1_Click(object sender, EventArgs e)
{
String Cust_id,Bail_account,Bail_total_amount,Bail_total_amount,Bail_status,FreezeFlag;
        Cust_id = DropDownList1.SelectedValue;
Bail_account = TextBox1.Text;
Bail_total_amount = TextBox2.Text;
Bail_status = DropDownList2.SelectedValue;
     FreezeFlag = DropDownList3.SelectedValue;
        if (IsValid)//如果当前页面通过验证
        {
            BailBLL pb = new BailBLL();
            Bool isok = pb.AddBail(Cust_id,Bail_account,Bail_total_amount,Bail_total_amount,Bail_status,FreezeFlag);
            if (isok)//如果信息添加成功,则跳转到保证金查看页面
                Response.Redirect("BailList.aspx");
        }
protected void Button1_Click(object sender, EventArgs e)
{
//teacherList 页面加载事件
        BailBLL pb = new BailBLL();//声明业务逻辑层对象,用于调用其中的方法
        GridView1.DataSource = pb.getBail();
        GridView1.DataBind();
    }
```

任务二十四：银行信贷管理系统的质押信息管理

一、任务描述

你作为《银行信贷管理系统》项目组的程序员，请实现如下功能：
- 查询质押信息；
- 编辑质押信息。

二、功能描述

（1）在图24.1所示的银行信贷管理系统主页面中，点击左侧导航菜单"客户质押信息管理"下的"查询/编辑质押信息"链接，进入"查看/编辑质押信息"页面，如图24.2所示。

图24.1　银行信贷管理系统主页面

（2）在图24.2中的文本框中输入要查询客户名称的匹配字符，点击"查询"按钮，在"查询/编辑质押信息"页面显示客户质押信息，如图24.3所示。

（3）点击选定行的"编辑"按钮，进入"编辑客户质押信息"页面，如图24.4所示。

（4）在图24.4中，修改质押信息，"＊"代表必填信息，点击"保存信息"按钮成功修改后跳转到"查询/编辑质押信息"页面。

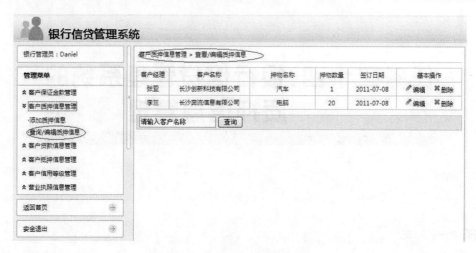

图 24.2 查看/编辑质押信息页面

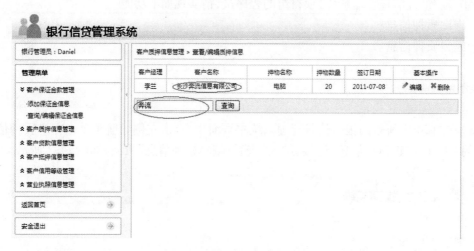

图 24.3 查询满足条件的客户质押信息

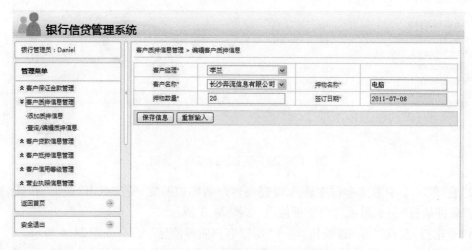

图 24.4 客户质押信息编辑页面

三、要求

1. 界面实现

以提供的素材为基础,实现图 24.1 ~ 图 24.4 所示页面。说明如下:

(1)在图 24.4 中,客户名称对应的下拉列表的 value 和 Text 属性用(2011070101,长沙创新科技有限公司),(2011070102,长沙奔流信息有限公司),(2011070103,长沙蓝海科技有限公司)三组值填充;

(2)在图 24.4 中,客户经理对应的下拉列表的 value 和 Text 属性用(001,张亚),(002,李兰),(003,王洪)三组值填充;

(3)在图 24.3 中,需将数据库中的客户经理编号和客户编号,根据下拉列表中的 value 和 Text 的对应关系,获取各编号所对应的 Text 值,显示在页面上。

2. 数据库实现

(1)创建数据库 BankCreditLoanDB。

(2)创建质押信息表 T_impawn_info,表结构见表 24.1。

表 24.1 质押信息表(T_impawn_info)表结构

字段名	字段说明	字段类型	是否允许为空	备注
Borrow_id	借据号	char(10)	否	主键
Cust_id	客户编号	char(10)	否	
Agency_id	经手人	char(3)	否	
Pawn_goods_name	押物名称	varchar(60)	否	
Pawn_goods_num	押物数量	int	否	
Contract_date	签定日期	datetime	否	

(3)在质押信息表 T_impawn_info 中插入记录,见表 24.2。

表 24.2 质押信息表(T_impawn_info)记录

Borrow_id	Cust_id	Agency_id	Pawn_goods_name	Pawn_goods_num	Contract_date
2011070801	2011070101	001	汽车	1	2011 - 07 - 08
2011070802	2011070102	002	电脑	20	2011 - 07 - 08

3. 功能实现

(1)功能需求如图 24.5 所示。

(2)依据活动图完成质押信息查询功能,如图 24.6 所示。

(3)依据活动图完成质押信息编辑功能,如图 24.7 所示。

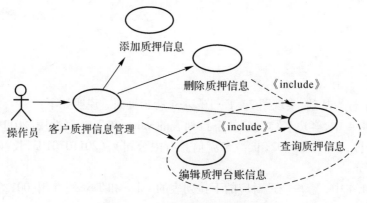

图 24.5　客户质押信息管理用例图

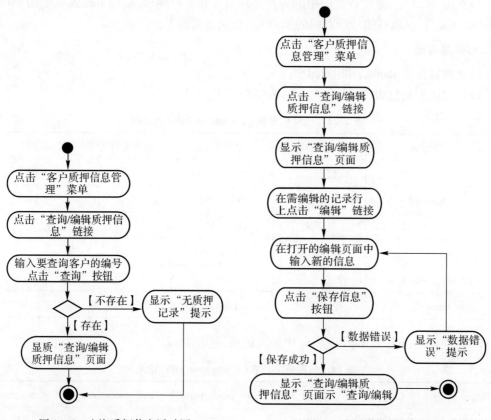

图 24.6　查询质押信息活动图　　　图 24.7　编辑质押信息活动图

四、必备知识

1. 数据库相关知识

（1）使用 MS SQL Server 2005/2008 创建数据库，创建数据表，设置表的字段，数据类型，主键，外键，约束。

（2）向数据表插入、删除、修改、查询数据。

2. 页面相关知识

（1）看懂系统所提供的素材，并理解框架技术。

（2）使用 asp.net 验证控件对页面必要的内容进行校验。

3. Asp.net 相关知识

（1）理解如何将 aspx 页面嵌入到框架中。

（2）使用数据源绑定控件显示数据库的数据。

（3）合理的使用转发和重定向控制项目的页面跳转。

（4）使用 ado.net 技术与数据库进行交互。

（5）Asp.net 三层结构的数据访问层、业务逻辑层和表示层的功能和合作

五、解题思路

1. 数据库思路

（1）根据项目要求创建数据库和数据表，向数据表中插入合适的测试数据。

（2）添加数据访问层，并在其中写好连接字符串和若干个通用的数据库操作方法。

（3）添加业务逻辑层，设置适当的参数，实现本任务所要求的全部功能，本层可以引用数据访问层的方法。

2. 表示层思路

（1）将提供的素材页面改写为 aspx 页面。

（2）将数据库数据的显示控件和部分表单控件改成 asp 服务器端控件。

3. 实现思路

（1）在表示层访问业务逻辑层，按照题目要求调用业务逻辑层的方法。

（2）数据添加的时候，正确获取用户在表单中提取的数据，并写入数据库中，然后进行页面跳转逻辑控制。

（3）查询的时候，调用业务逻辑层的方法，获取数据库中的数据，并按照用户要求的格式显示出来。

六、操作步骤

1. 业务逻辑层代码

```
public class T_impawn_infoBLL
{
    //按客户名称查询全部或者部分质押信息;
    public DataSet getimpawn(string key)
    {
        string sql = string.Format("select Borrow_id,Agency_name,Cust_name,Pawn_goods_name,Pawn_goods_num,Contract_date from T_impawn_info t,Cust c ,Agency a where a.Agency_id = t.Agency_id and c.Cust_id = t.Cust_id and Cust_name like '%{0}%'", key);
```

```csharp
        DBHelper db = new DBHelper();
        return db.FillDataSet(sql);
    }
    //编辑质押信息。
    public int updateimpawn(string Cust_id,string Agency_id,string Pawn_goods_name,string Pawn_goods_num,string Contract_date,string id)
    {
        string sql = string.Format("update T_impawn_info set Cust_id='{0}',Agency_id='{1}',Pawn_goods_name='{2}',Pawn_goods_num={3},Contract_date='{4}'where Borrow_id='{5}'",Cust_id,Agency_id,Pawn_goods_name,Pawn_goods_num,Contract_date,id);
        DBHelper db = new DBHelper();
        return db.Operate(sql);
    }
    //根据Borrow_id获取当前质押信息
    public DataSet getimpawnbyBorrow_id(string id)
    {
        string sql = string.Format("select * from T_impawn_info where Borrow_id={0}",id);
        DBHelper db = new DBHelper();
        return db.FillDataSet(sql);
    }
    //获取所有的客户经理信息
    public DataSet getallAgency()
    {
        string sql = "select * from Agency";
        DBHelper db = new DBHelper();
        return db.FillDataSet(sql);
    }
    //获取所有的客户名称信息
    public DataSet getallCust()
    {
        string sql = "select * from Cust";
        DBHelper db = new DBHelper();
        return db.FillDataSet(sql);
    }
}
```

2. 表示层代码

```csharp
public partial class ImpawnList : System.Web.UI.Page
{
    protected void Page_Load(object sender, EventArgs e)
    {//页面首次加载的时候,查询所有的客户质押信息
        if (!IsPostBack)
        {
```

```csharp
            T_impawn_infoBLL tb = new T_impawn_infoBLL();
            GridView1.DataSource = tb.getimpawn("");
            GridView1.DataBind();
        }
    }
    protected void Button1_Click(object sender, EventArgs e)
    {//根据关键字搜索客户质押信息
        T_impawn_infoBLL tb = new T_impawn_infoBLL();
        GridView1.DataSource = tb.getimpawn(TextBox1.Text);
        GridView1.DataBind();
    }
    protected void LinkButton2_Command(object sender, CommandEventArgs e)
    {
        string id = e.CommandArgument.ToString();
        Response.Redirect("ImpawnEdit.aspx?id=" + id);
    }
}
public partial class ImpawnEdit : System.Web.UI.Page
{
    static string id;
    protected void Page_Load(object sender, EventArgs e)
    {//页面加载的时候,判断是否有查询字符串传过来,如果没有,则自动跳转到客户质押信息查询页面
        if (Request.QueryString["id"] == null || Request.QueryString["id"].ToString() == "")
            Response.Redirect("ImpawnList.aspx");
        else
        if (!IsPostBack)
        {
            T_impawn_infoBLL tb = new T_impawn_infoBLL();
            DropDownList1.DataSource = tb.getallCust();
            DropDownList1.DataTextField = "Cust_name";
            DropDownList1.DataValueField = "Cust_id";
            DropDownList1.DataBind();
            DropDownList2.DataSource = tb.getallAgency();
            DropDownList2.DataTextField = "Agency_name";
            DropDownList2.DataValueField = "Agency_id";
            DropDownList2.DataBind();
            id = Request.QueryString["id"].ToString();
            DataSet ds = tb.getimpawnbyBorrow_id(id);
            if (ds.Tables[0].Rows.Count > 0)
            {
                DropDownList1.SelectedValue = ds.Tables[0].Rows[0]["Cust_id"].ToString();
                DropDownList2.SelectedValue = ds.Tables[0].Rows[0]["Agency_id"].ToString();
```

```
                TextBox1.Text = ds.Tables[0].Rows[0]["Pawn_goods_num"].ToString();
                TextBox2.Text = ds.Tables[0].Rows[0]["Pawn_goods_name"].ToString();
                string aa = ds.Tables[0].Rows[0]["Contract_date"].ToString();
                int index = aa.IndexOf(' ');
                TextBox3.Text = aa.Substring(0, index);
            }
        }
    }
    protected void Button1_Click(object sender, EventArgs e)
    {//保存修改
        string Cust_id, Agency_id, Pawn_goods_name, Pawn_goods_num, Contract_date;
        Cust_id = DropDownList1.SelectedValue;
        Agency_id = DropDownList2.SelectedValue;
        Pawn_goods_num = TextBox1.Text.Trim();
        Pawn_goods_name = TextBox2.Text.Trim();
        Contract_date = TextBox3.Text;
        T_impawn_infoBLL tb = new T_impawn_infoBLL();
        if (tb.updateimpawn(Cust_id, Agency_id, Pawn_goods_name, Pawn_goods_num, Contract_date, id) > 0)
            Response.Redirect("ImpawnList.aspx");
    }
}
```